Atanásio Raimundo Massingue

Socio-Environmental Vulnerability to Water Scarcity

Atanásio Raimundo Massingue

Socio-Environmental Vulnerability to Water Scarcity

Case of Funhalouro District Inhambane Province

ScienciaScripts

Imprint

Any brand names and product names mentioned in this book are subject to trademark, brand or patent protection and are trademarks or registered trademarks of their respective holders. The use of brand names, product names, common names, trade names, product descriptions etc. even without a particular marking in this work is in no way to be construed to mean that such names may be regarded as unrestricted in respect of trademark and brand protection legislation and could thus be used by anyone.

Cover image: www.ingimage.com

This book is a translation from the original published under ISBN 978-620-2-56092-4.

Publisher:
Sciencia Scripts
is a trademark of
Dodo Books Indian Ocean Ltd., member of the OmniScriptum S.R.L Publishing group
str. A.Russo 15, of. 61, Chisinau-2068, Republic of Moldova Europe
Printed at: see last page
ISBN: 978-620-2-70317-8

Index

List of Abbreviations 8

Dedication 11

Acknowledgements 12

Summary 13

Introduction 15

Specific Objectives 19

Problematization 19

Justification 20

1.0 Conceptual and Theoretical Framework of Vulnerability. 28

1.1 Elements of a Risk 29

    1.1.2 Exposure 29

    1.1.3 Danger 30

    1.1.4 Hazard Analysis 30

    1.1.5 Risk Analysis 30

1.2 Vulnerability Calculations 32

1.4 Vulnerability Analysis 34

1.5 Vulnerability Analysis as Damage Exposure 34

1.6 Social Vulnerability 35

1.6 Environmental Vulnerability 36

1.7 Water Resources 36

1.8 Hydrological Principles of Water Resources 37

1.9 Hydrological Cycle Processes 37

    1.9.1 Precipitation 38

    1.9.2 Interception 38

    1.9.3 Plant interception 38

    1.9.4 Storage in Depressions 39

1.9.5 Infiltration 39

1.9.6 Evaporation 40

1.9.7 Surface Flow 40

1.9.8 Coefficient of Surface Flow (C) 40

1.10 Freshwater Distribution Worldwide... 40

1.11 Water Resources Uses 41

1.11.1 Fresh water consumption in developed countries....................................... 41

1.11.2 Developing Countries 41

1.12 The Difference Between Water Scarcity and Drought...................................... 41

1.13 Causes and Consequences of Water Scarcity... 42

1.13.1 Aridity... 43

1.13.2 Drought.. 43

1.13.3 Desertification 44

1.14 Social Context of Water Scarcity 44

1.15 Consequences of Water Scarcity.. 45

1.16 Societal Environmental Value of Water... 45

1.17 The Global Water Scarcity Situation 46

1.17.1 Australia 47

1.17.2 Kenya 47

1.17.3 Sahel Region 47

1.17.4 Tanzania 47

1.17.5 Mozambique 48

1.18 Water Scarcity Adaptation Strategies in Some Countries of the World......... 48

1.18.1 Brazil... 48

1.18.2 South Africa... 49

1.18.3 Namibia.. 49

1.18.4 Kenya 49

1.18.5 Australia 49

1.18.6 Mozambique 50

1.19 Risk and Vulnerability Situation of Mozambique in Water Scarce 50

1.20 International Water Conventions and Protocols.. 51

1.20.1 SADC Protocol on Shared Water Courses 52

1.20.2 United Nations Conventions 52

1.20.3 Water Scarcity Conventions (The Hague Declaration) 53

1.20.4 Declaration on Water and Sustainable Development 53

1.20.5 Interim Tripartite Agreement between the Republic of Mozambique, the Republic of South Africa and the Kingdom of Swaziland (Resolution n° 53/2004 of 1 December 2004) 54

1.21 Sustainable Development Goals (SDOs) 54

1.22 National Water Resources Management Strategy 55

1.23 Water Supply in Rural Areas 55

1.24 Sanitation 56

1.25 Water Resources and Climate Change in Mozambique 56

1.26 Relationship of El Nino Phenomenon, South Oscillation (ENSO) to Water Scarcity 57

1.27.1 Impact of Enso on the World 57

1.27.2 Impact of Enso on Southern Africa 58

1.28 Impact of Climate Change on Water Resources 59

1.29 Methodologies for Socio-Environmental Vulnerability Analysis at Water Scarcity 61

1.29.1 Falkenmark Water Stress Index (WSI) 61

1.29.2 Social Water Stress Index 62

1.29.3 Basic Human Water Requirements 62

1.29.4 Water Poverty Index (WPI) 62

1.29.5 Water Footprint 63

1.29.6 Virtual Water 63

1.29.7 Physical and Economic Water Scarcity Indicator (IWMI) 64

1.29.8 Index Based on Water Withdrawal 65

1.29.9 Water Supply Sustainability Risk Index 65

1.29.10 Climate Index 66

1.29.11 Vulnerability Index on the basis of the Declivities 67

1.29.12 Representative Table of Values of Flow Coefficients according to Soil Type, Declivity and Plant Cover. 67

1.29.13 Vulnerability Index on the Basis of Soil Type Characteristics 68

1.29.14 Vulnerability Index at the Vegetation Base 69

2nd Chapter 69

2.0 Physico-Geographical and Socio-Economic peculiarities of Funhalouro District
.... 69

2.1 Location of Funhalouro District.. 69

2.2 Administrative Division of Funhalouro District.................................... 70

2.3 Geology 70

2.4 Relief 70

2.5 Soil... 70

    2.5.1 Sandy and Dune Roof Soils.. 71

    2.5.2 Mananga Sediments.. 71

2.6 Hydrography... 72

2.7 Climate... 73

2.8 Vegetation.. 74

2.9 Socio-Economic Aspect of Funhalouro District.................................. 75

    2.9.1 Population.. 75

2.10 Historical Overview of District Name.. 76

    2.10.1 Religion 76

2.11 Socio-economic Infrastructures 76

2.12 Socio-Economic Activities.. 79

III Chapter: Presentation and Analysis of Water Scarcity Response Data in the District of Funhalouro... 80

    Falkenmark Water Stress Index (WSI) 81

    Water Footprint 81

    Virtual Water... 81

3.1 Environmental Vulnerability to Water Scarcity 81

    3.1.1 Lithological Factor.. 81

    3.1.2 Factor of Declivity.. 85

    3.1.3 Soil Salinisation Factor... 87

    3.1.4 Climate Factor... 89

3.1.5 Vegetation Factor 90

3.2 Social Factor 93

3.3 Institutional Factor 97

3.4 Hazards of Water Vulnerability in Funhalouro District 99

Conclusion 99

Recommendations 101

Bibliography 102

List of Abbreviations

Abbreviation		Meaning
AIM	-	Mozambique Information Agency
ANA	-	National Water Agency
ARA - SOUTH	-	Regional Water Administration - South
BCI	-	Commercial Investment Bank
CPTEC	-	Weather Forecasting and Climate Studies Centre
DNA	-	National Directorate of Water
DPOHPI	-	Provincial Directorate of Public Works and Housing of Inhambane
EEA	-	European Environment Agency
EPC	-	Complete Elementary School
ESF	-	Enginyeria Sense Fronteres
ESG	-	General High School
FAO	-	Food and Agriculture Organization of the United Nations
FPA	-	Food and Agriculture Organization of the United Nations
GF	-	Government of Funhalouro
IBGE	-	Brazilian Institute of Geography and Statistics
INAM	-	National Institute of Meteorology
INE	-	Statistics Portugal - National Statistical Institute
INGC	-	National Institute of Disaster Management
IPCC	-	Intergovernmental Panel on Climate Change
IWMI	-	International Water Management Institute
MAE	-	Ministry of State Administration
MAMAOT	-	Ministry of Agriculture, Sea, Environment and Land Management.
MCEL	-	Mozambique Mobile
MICOA	-	Ministry for Environmental Action Coordination
MISAU	-	Ministry of Health
MDG	-	MDG - Sustainable Development Goal
ODS	-	Sustainable Development Objective
OECD	-	Organisation for Economic Cooperation and Development
NGO	-	Non-governmental organizations

UN	-	United Nations
PEDD	-	Strategic District Development Plan
WEIGHT	-	Strategic Plan for Water and Rural Sanitation
ASR		
PMA	-	WFP - World Food Programme
PNA	-	National Water Policy
UNDP	-	United Nations Development Programme
PQG	-	Government Five Year Programme
PRONASAR	-	National Program of Water Supply and Rural Sanitation
SADC	-	Southern African Development Community
SDPI	-	District Planning and Infrastructure Service
GIS	-	Geographical Research System
TSM	-	Surface Sea Temperature
UNDP	-	United Nations Development Programme
UNECE	-	United Nations Economic Commission for Europe
UNESCO	-	United Nations Educational, Scientific and Cultural Organization
UNICEF	-	United Nations Children's Fund
UN-ISDR	-	United Nations- International Strategy for Disaster Reduction
UN WATER	-	United Nations Water
UTM-	-	Universal Transverse Mercator
WWDR		World Water Development Report

Declaration

I declare that this dissertation is the result of my personal investigation and the guidance of my supervisor, its content is original and all sources consulted are duly mentioned in the text, notes and final bibliography.

I also declare that this work has not been presented in any other institution for any academic degree.

Maputo______/______/______

The student

____

Atanasio Raimundo Massingue

Dedication

I dedicate this work to my son Raimundo Atanásio Massingue, my first luck.

Acknowledgements

It is with great honour and joy that I address my thanks to my supervisor Prof. Dr. Sabil Damião Mandala, who tirelessly guided me in the activities to carry out this work, responding to all the concerns that arose throughout the execution of this activity and above all giving motivation to achieve the goals we wanted.

To my colleagues in the environmental management course, especially Alcídio Muhatuque, Feniasse Michaque and Marcelino Mondlane, I would like to express my gratitude for the support in ideals that contributed to the materialization of this dissertation.

In memory of my uncle Sérgio Carlos Massingue I thank you for the reinforcement of the stick that you gave me in primary school

My thanks also go to Master Apolinário Joaquim Malauene Nhamposse who, from high school to college, aroused in me the curiosity, interest and desire to acquire knowledge of the geographical sciences.

I would also like to thank Prof. José Julião da Silva, director of the Environmental Management course, for supporting my application to the master's degree course in environmental management.

To Prof. Gustavo Sobrinho Djedje, (Director of the Faculty of Earth Science and Environment) who in his discipline of risk management has awakened in me the desire to apply through this study the matters of environmental risk also goes my thanks.

My thanks also go to teachers Zacarias Ombe, António Suluda and Suzete Buque.

My very Khanimambo also goes to the National Water Directorate (DNA), for providing the hydrogeological map of Mozambique, to the Agrarian Research Institute of Mozambique (IIAM) for providing the soil map of Mozambique and to the SDPI of Funhalouro for providing information on this research.

To the population of Funhalouro district, "Ndzi bonguile nguvu" (thank you very much) for the information, instructions and teaching they have given me in the forms of water management.

Finally, I would like to thank everyone who directly or indirectly contributed to this study of the dissertation on environmental management.

12

Summary

The district of Funhalouro, being located in a semi-arid area, the search for strategies for the mitigation of water scarcity is a priority action in order to reduce the great distances that the population travels in the search for water and the risk of food insecurity as well as the frequent losses of livestock, in this context this research aims to analyze the capacity of response to water scarcity used as a method the International Water Management Index (IWMI). The methodological procedures for information processing and thematic maps were carried out using the Geographic Information System *software*, QGIS, and ArcGIS. The physical components of the study area were inferred using cartographic data (topographic maps 1: 250,000) where field work (observation) and interview information were integrated. This research concludes that water scarcity in this district is the result of a combination of natural and human factors and that among the natural factors causing water scarcity, lithology due to the existence of salt capes left during the period of marine regression thus contributing to the occurrence of water with higher sodium chloride content in several places of the district. This fact is further aggravated by environmental degradation and climate changes that are reflected in the decrease of precipitation and increase in temperature accelerating the processes of soil salinization. With regard to human factors, despite the implementation of actions such as rainwater harvesting and storage in cisterns, the construction of dams and small water treatment centres note that the lack of financial resources and qualified human capital limits the actions of expansion of the water supply network and maintenance of infrastructure, which happens in systematic failures of water harvesting equipment and subsequently its inoperation.

Keywords: Socio-Environmental Vulnerability to Water Scarcity

Abstract

As Funhalouro district is located in a semi-arid zone, the search for strategy to mitigate water scarcity is a priority action in order to reduce the large distance that the population travel in search of water and the risk of food insecurity as well as frequent loss of cattle. In this context, the present research aims to analyze the capacity to respond to water scarcity using the international water management index (IMWI) as a method. The methodology procedure for the treatment of information and the preparation of the thematic maps were performed using the geographic information system software, QGIS and ArcGIS. The physical components of the study area were inferred using cartographic map 1: 250 000, where fieldwork observation and interview information were integrated. The present research concluded that the water scarcity in this district is a result of the combination of natural and human factors, and among the natural factors that cause the water scarcity, the lithology is highlighted due to the existence of salt layer left during the marine regression period what contribute the occurrence of higher sodium chloride water in various part of the district. This fact is further aggravated by environmental degradation and climate change, which is reflected in the decrease of precipitation and temperature increase, accelerating the processes of soil salinization. Regarding human factors, despite the implementation of action such as the capture of rainwater and its storage in cistern, the construction of dams and water treatment center, is noted that the lack of financial resource and qualified human capital limit the actions of enlargement of the water supply network and maintenance of infrastructure breakdowns of the water catchment equipment and subsequently its inoperation.

Key Word: Social and Environmental, water scarcity vulnerability

Introduction

The present work entitled " Socio-environmental vulnerability to water scarcity - Case of the District of Funhalouro Inhambane Province, arises from the perspective of acquiring the degree of Master in Environmental Management by the Pedagogical University of Maputo, seeking the experiences adopted by the population in the management of water resources in the face of the harsh reality of its scarcity.

Water is an indispensable natural resource for the survival of man and other living beings and is a fundamental substance for the ecosystems of nature, as well as for atmospheric water formations, influencing the climate of the regions, however, this natural resource is being exhausted by the actions of man having as consequences the degradation of its quality, the reduction of its ecosystem service and its supply value due to its exhaustion. The concern for water scarcity brought together several world-class experts in the city of Toronto, Canada, in March 2011 to discuss the global stage of drinking water supply as part of global security issues since water scarcity has already reached very dangerous levels affecting 26 countries such as Saudi Arabia, Iraq, Belgium, Mexico, Ethiopia, France, Egypt, Thailand, the United States of America, Burundi, among others, and this problem is for many countries one of the limiting factors for development.

In addition to the countries mentioned there are also forecasts that in the coming years people may live in regions with extreme water shortages, including for consumption, but the reason for this warning is debatable among water resource researchers, for example other authors combine this problem with natural factors (dependence on climate and the physical and biological characteristics of the ecosystems that comprise it), others look at the problem from the point of view of supply and demand, while others claim that the problem is much more of a management than a real crisis of scarcity and stress.

This divergence in identifying water stress is justified by the multiple important facets of water use, supply and demand, which has led to the creation and selection of criteria by which water problems are assessed in order to implement correct political or scientific decisions.

The introduction of water scarcity indices and assessment methodologies culminated in the classification of countries with physical water scarcity problems and others with economic water scarcity problems.

In the case of Mozambique, the use of water scarcity assessment methodologies has classified as the country with physical and economic water scarcity problems and the reasons for water scarcity are the deficiency in its distribution, unequal occurrence and availability of water resources, as well as the dependence on rainfall

For the preparation of this dissertation it was necessary to follow the work steps described below:

1st Chapter: Conceptual and Theoretical Framework of Vulnerability

A bibliographic survey of the subject under study was carried out and its revision resulted in the conceptualization of socio-environmental vulnerability and its elements, description of the hydrological principles of water resources, use of water resources in developed and developing countries, international conventions on water, water resources and climate change in Mozambique and methodologies for analysis of socio-environmental vulnerability to water scarcity.

2nd Chapter: Socioeconomic Particularities of the Study Area

In this chapter we continued with the bibliographic review and the consultation of topographic maps of the physical and geographical aspects of the study area in order to represent the phenomena analysed in the thematic map through the Qgis and Arcgis software. This research resulted in the physical and geographical characterization of the district of Funhalouro as well as in the discretion of the socioeconomic infrastructures of the district of Funhalouro.

3rd Chapter: Physical-Geographical and Socio-Economic Particularities of Funhalouro District

In this chapter, based on the data from the first and second chapters, as well as the data obtained during the interview and questionnaire, an analysis of the response capacity to water scarcity in the study area was made. This analysis is reinforced through the use of figures, tables, charts or graphs duly subtitled to facilitate the understanding of the subjects.

Conclusion

In a nutshell, this chapter presents the factors of vulnerability to water scarcity in Funhalouro district and the actions that are being taken to mitigate the effects of water scarcity.

Recommendations

Finally, some long and short term recommendations are proposed in order to improve the capacity to respond to water scarcity in the district.

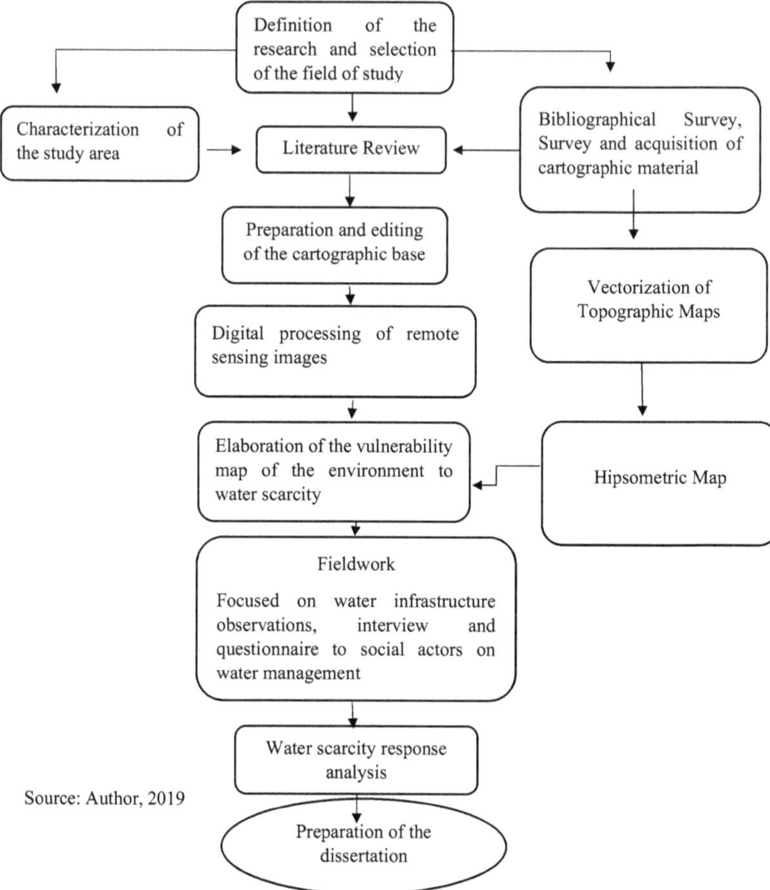

Source: Author, 2019

General Objective

✓ Analyze Water Scarcity Response in Funhalouro District.

Specific Objectives

✓ Describe the physical-geographical particularities of the district of Funhalouro
✓ Describe the socioeconomic particularities of Funhalouro district
✓ Demonstrate the factors of water scarcity in the district of Funhalouro
✓ Relate the actions implemented to make water available to the water needs of the district of Funhalouro

Problematization

In 2014 until the beginning of 2016 the water shortage in the district of Funhalouro has put about 30 thousand people in a situation of food insecurity and the death of several faunal species, such as cattle and goats. For this scenario the population was forced to resort to roots and wild fruits for their food and to travel more than 20 km in search of water to meet their needs. Knowing that the district of Funhalouro because it is located in a semi-arid area is vulnerable to the occurrence of water scarcity - **What strategies does the population adopt to mitigate the effects of water scarcity?**

Hypothesis:

Vulnerability to water scarcity in the district of Funhalouro is a product of spatial, institutional and social contexts that require structural and non-structural measures such as the development of rainwater harvesting and storage techniques, construction and maintenance of water infrastructures as well as systematic actions to preserve the environment as a way to face global climate change.

Justification

As we already know, water is a vital environmental resource for the permanence of human beings on planet earth besides being responsible for the balance of ecosystems, however, its scarcity puts at risk the provision and ecosystem services, thus placing the need for sustainable use based on the elaboration and implementation of appropriate management strategies for this resource for the maintenance of life on earth.

The southern part of Mozambique, for example the district of Funhalouro, is highly vulnerable to extreme events such as the drought that is most felt in water scarcity and agricultural activity putting thousands of people in a situation of food insecurity. This phenomenon has become a major concern for government, civil society and non-governmental organizations (NGOs) in finding ways to mitigate the impacts of climate change in general and water scarcity in particular.

In the search for solutions, this study seeks experiences and their limitations in the management of water resources implemented in the semi-arid area of Funhalouro district in order to be rethought and implemented according to the geographical provisions and social reality of this district as well as the implementation of these strategies in other districts such as Moamba in Maputo province, Chigubo and Chicualala in Gaza province, just to cite a few examples with the aim of reducing the suffering of the population in the search for water.

Knowing that the water problem in Funhalouro is mostly economic which makes it impossible to solve the problems in the long term there is a need to revitalize the thought of the comparative advantages of the district as an opportunity to overcome the challenges of water scarcity now let us see the exploitation and marketing of forest resources of great value marginalises the population and this would be the added value that would reduce the economic problems in providing water to the population.

Methodology

This study takes a qualitative approach although it uses the quantitative method to make the analysis of the socio-environmental vulnerability to water scarcity in the scope of the capacity of response in the district of Funhalouro.

Thus, the methodological procedure was based on the literature review for the theoretical basis and documentary research through the collection of secondary data from official sources as detailed below.

Literature Review

This method consisted of consulting books, magazines, dissertations and theses and publications available in virtual databases, electronic libraries as well as reading national press news papers that report the problem of water scarcity in the study area.

For the development of this research the following main bibliographical references were adopted: Mandala (2016), Almeida (2010), Rezende (2016) Kobiyama (2008), Raskin (1997), Weng (2016), Dos Muchangos (1997), Micoa (2005), FAO (2011), water law 1991,

These bibliographical references were consulted in the library of the Pedagogical University of Maputo, Brazão Mazula library, the library of the National Water Directorate as well as in the database of the Capes Portal, *Scientific Electronic Library Online* (Scielo); Biofund Virtual Library, riverawarenesskit

Documental Research

According to Andrade & Lakatos (2007:176) documental research is restricted to written or unwritten documents, constituting what is called primary sources, such as primary documents, public archive documents, parliamentary publications, travel reports, photographs, maps, etc.

However, this method consisted of collecting national and international primary sources

a) National Sources

- ✓ The National Water Policy, August 2007
- ✓ The water law of 16/91 of 3 August 1991

b) International Sources

✓ International agreement ratified by **resolution n° 53/2004 of 1 December.**

✓ The SADC protocol on ractified watercourse sharing in 2003.

The data collected also served as a collection of data for the formulation of an interview script and questionnaire in order to make up for the inadequacy of the information under study.

Cartographic Base

This method was used to obtain Declivity, Soil and Vegetation data by geoprocessing[1] topographic charts on the scale of 1:250,000 represented in figure 2, using tools from the Quantum GIS software (Qgis).

Figure 1: Geo tiff format map scanner

Source: Cenacarta

Based on the georeferenced data, the hypsometric map was drawn up with the following phases:

1. Georeferenced the above letters in Mercator Transverse Universal Meridian (UTM) central 33° E projection of Greenwich Meridian, ellipsoid of Clark, 1866 with the re-edition dated 2002, the equidistance of the contour lines is 100 meters. The following table shows the control points and the margin of error.

Table 1: Control points of georeferenced charts

Topographic charts	Control points	Error in (degrees)

---

[1] According to Rezende apude Camara (2016:27) geoprocessing uses software and mathematical techniques for the treatment of geographic information and from this operation allows to spatialize and perform complex analysis by integrating several variables. In view of this, it has several systems such as the geographic information system - GIS.

	Latitude (φ)	Longitude (λ)	
Chigubo	22°00′	33°00′	0,00184741
	22°00′	34°00′	
Letter $\frac{SF\text{-}36}{P}$ Sheet No. 85	23°00′	34°00′	
	23°00′	33°00′	
Funhalouro	23°00′	34°00′	0,001777
	23°00′	35°00′	
Letter $\frac{SF\text{-}36}{X}$ Sheet No. 91	24°00′	35°00′	
	24°00′	34°00′	
Changane River	23°00′	33°00′	0,00170066
	23°00′	34°00′	
Letter $\frac{SF\text{-}36}{V}$ Sheet No. 90	24°00′	34°00′	
	24°00′	33°00′	
Mabote	22°00′	34°00′	0,00182378
	22°00′	35°00′	
Letter $\frac{SF\text{-}36}{Q}$ Sheet No. 86	23°00′	35°00′	
	23°00′	34°00′	
Mapinhane	22°00′	35°00′	0,00167797
	22°00′	36°00′	
Letter $\frac{SF\text{-}36}{R}$ Page 87,	23°00′	36°00′	
	23°00′	35°00′	

Source: Author's elaboration based on Cenacarta maps data.

2. After the georeferencing of the charts the mosaic of the charts was made with a buffer of 100 meters in the limits of the district as well as the vectorization of the contours of levels with an equidistance of 100 meters and points quoted in order to obtain the relief of the study area as illustrated in the figures below.

Figure 2: Mosaic of Georeferenced Letters

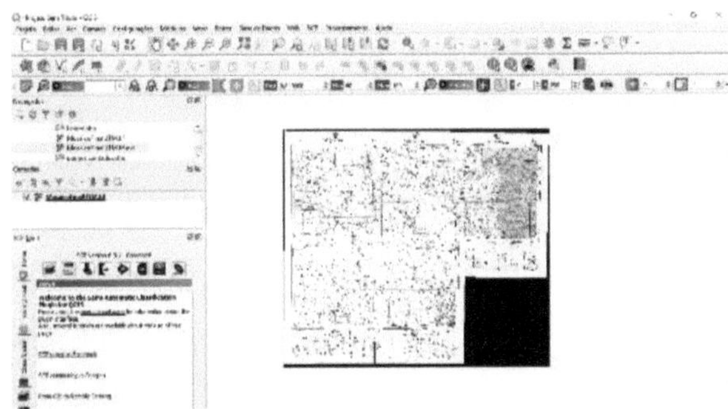

Source: Author's elaboration based on CENACARTA maps

Figure 3: Representation of the 100 meter Buffer at the district boundaries.

Figure 4: Representation of the vectorization of the contour lines

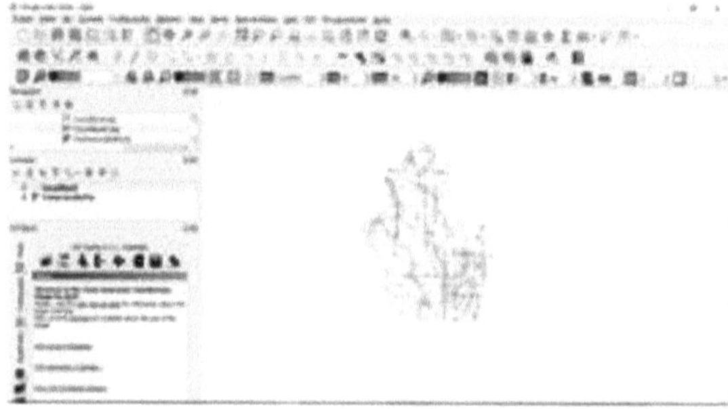

Figure 5: Vectorisation of lakes

Source: Author's elaboration based on CENACARTA maps

Direct Note

In the direct observation, the credential was first acquired by the Pedagogical University of Maputo, see Appendix No. 2, which was presented in the structures of the government of the district of Funhalouro and the heads of localities of the district, in order to authorize the realization of this research. The observation was accompanied by data collection instruments, namely: the questionnaire, the interview and the observation guide presented below, which was based on the environmental, social and institutional component. A Tecno Mobile camera with a 13 mega picture resolution was used as an instrument to aid direct observation for the recording of images that helped in the analysis of this study.

Figure 6: Observation Scheme

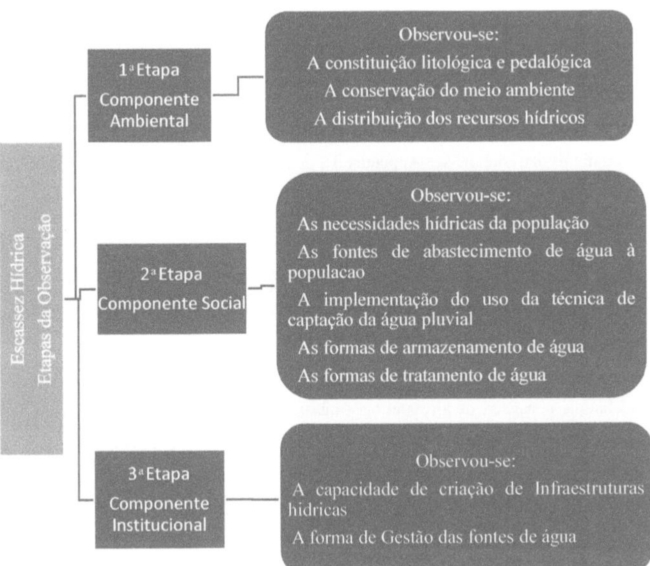

Source: Author's elaboration, 2019

Interview

According to Andrade & Lakatos (2007:197) the interview is a procedure used in social research to collect data or to help in the diagnosis or treatment of a social problem.

This method consisted of having information from four (4) persons with recognized authority in the district as illustrated in the chart below:

Nº	Identification	Code	Interview location
1	SDPI representative	E. F. 1	Funhalouro Headquarters
1	Private water distribution operator	E. F. 2	Mucuine
1	Health Technician	E. F. 3	Funhalouro Headquarters
1	Professor of Geography	E. F. 4	Tsename

For various reasons the interview lasted 9 days, from 05/07/2019 to 14/07/19 and was conducted in Portuguese language and the information was noted and subsequently presented in this dissertation, see Appendix 3

Questionnaire

According to Gil (2006: 121) it is a research technique composed of a set of questions that are submitted to a person with the purpose of obtaining information about knowledge, beliefs, feelings, values, interests, expectations, aspirations, fears, present or past behaviors, etc.

This technique consisted of formulating a questionnaire to gather information on the strategies adopted to adapt to water scarcity in Funhalouro district, see Appendix 4.

For this survey, which lasted four (4) days, a sample of one hundred and eighty (180) inhabitants from various social strata, who are at least residents or workers in the district, divided into 90 inhabitants for each Administrative Post in the district, was selected between 08/03/2019 and 08/07/2019.

The questionnaire was prepared in Portuguese but its interpretation to the surveyed population was done in the local language of the study area (Xitswa) since most of the population has no command of the Portuguese language. The survey answers were recorded in the same survey which contained 4 open questions and 11 closed questions.

The compilation of the survey results is presented in this dissertation in the form of tables and graphs in order to facilitate the analysis of the data.

Analytical

According to Limon (2007) The analytical method consists in breaking down a phenomenological process as a whole into parts so that one can understand the elements as a whole which are prone to the observation of causes, the nature of the phenomena and their effects, thus enabling the establishment of new theories,

This method consisted in making interpretations of data obtained by cartographic methods, interview and survey with subjects related to the subject under study and this interpretation is represented with the descriptions, deductions, opinions, conclusions on the reality studied so that one can understand the elements in its whole daring of the observation of causes, the nature of phenomena and their effects, thus enabling the establishment of new theories.

Statistical or Mathematical Method

According to Andrade and Lakato (2007:93) the method means the reduction of sociological, political, economic phenomena, etc., in order to have quantitative data and statistical manipulation that makes it possible to prove the relationships between the phenomena and obtain generalizations about their nature, occurrence or meaning.

This method took place in August until the conclusion of the present dissertation through quantitative graphical representations of the phenomenon under study.

1st Chapter

1.0 Conceptual and Theoretical Framework of Vulnerability.

It is pertinent to note that the notion of vulnerability is intrinsically associated with risk although it differs from risk in its conceptualization. According to Castro et al (2005:11) it is admitted that the first technical studies on risk and uncertainty appeared in the literature in 1921 through the classic work titled "*Risk, uncertenity and* **profit**", by Frank Knight, who stated: "*if you do not know for sure what will happen, but the chances exist, that is risk. If you don't know what the chances are, then that's uncertainty"*. The **etymology of the term "Risco"** is little known, but for some authors like Nardocci, (1999:13) it is admitted that the word is of maritime origin where the edges of cliffs, reefs represented danger to sailors, and for other authors like Franklin (1998:1) it is admitted that this term was introduced in England in the seventeenth century in the context of gambling, as well as in statistics as the probability of a particular adverse event happening during a given period of time, or resulting from a particular challenge. It is further stated that this term has its origin in the Angliciazação de *risqué* which in French means offense. According to Veyet, (2007: 13) the conception and concern about risk in human history was not the same, because in the past man was more susceptible to disasters, famine, epidemics, which were perceived as divine punishments, but today modern societies are more vulnerable to meteorological fluctuations that translate into effects on society, having as causes not only natural but also human and technological as illustrated below

Figure 7: Diagram of risk types.

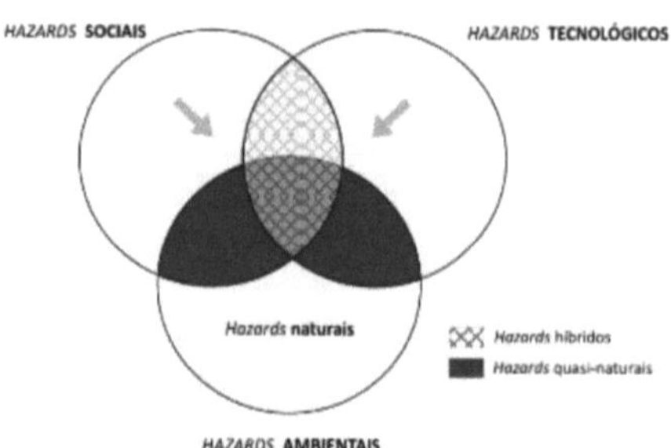

Source: Esteves, 2011: 68

In this context looking at the dimensions of risk, according to Nardocci (1999):12) it is not easy to define it as it seems because there are many concepts and different definitions in use in the most diverse areas of current science such as psychology, economics, medical and biological sciences, engineering and statistics, but nevertheless in general terms According to UN-ISDR, (2004) Risk can be understood as *the "probability of harmful consequences or expected losses (deaths, injury, property, livelihood, economic activity, disruption or damaged environment) resulting from the interaction between natural or human-induced hazards and vulnerable conditions"* i.e. the risk appears to be associated with the natural environment and physical phenomena present in study on potential, fragility, susceptibility, vulnerability, sensitivity or damage (SOUZA, 2015:34).

In short, risk involves exposure to external hazards over which people have limited control, while vulnerability measures the ability to combat such hazards without suffering a potential loss of well-being in the long term.

1.1 Elements of a Risk

Risk as a probability or as an indicator of the uncertainty of the occurrence of a harmful event has as its element: **Exposure, Danger and Vulnerability**, in a schematic form, these elements can be represented according to figure 8.

Figure 8: Model of the representation of the Risk components

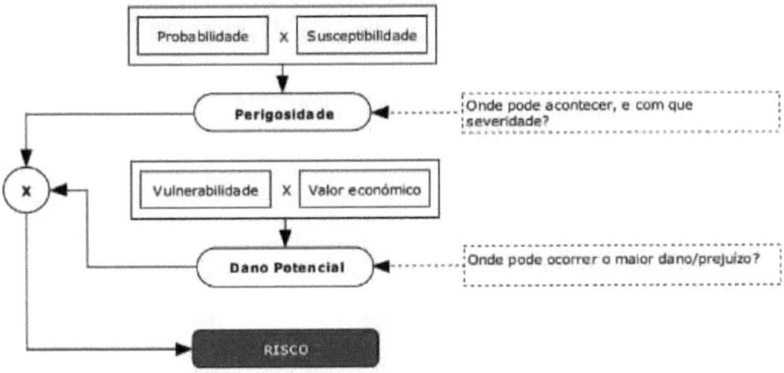

Source: Green, (2008:14)

1.1.2 Exposure

Exposure according to Cardona (2001:14) is understood as the condition of susceptibility or propensity of a given sand or territorial unit to be affected by the phenomenon, in this case of water shortage.

1.1.3 Danger

According to Almeida (2010:15) Danger is a potential threat to people and their property as opposed to risk as a probability of a hazard occurring and generating losses

1.1.4 Hazard Analysis

According to OECD, (2012:31) The hazard can be described in terms of physical phenomenon, probability and frequency of occurrence, location and direction of its process, intensity and scale of its hazard and its duration.

1.1.5 Risk Analysis

According to VIANA (2010:16) risk analysis *"is the study aimed at identifying the danger of an activity, project or area, also incorporating risk assessment"* and risk assessment is defined as a process of estimating the existing risk for possible recipients, which can be either goods, people or the environment, and management measures should be proposed, both preventive and emergency actions in a possible accident in order to reduce the risk and minimize the adverse consequences of damage (UNDP, 2005:1). Risk analysis also considers the nature and extent of a given risk through hazard analysis and assessment of existing conditions of vulnerability that together can harm people, property, services, and the environment exposed.

According to De Amaral, (2012:463) There are several techniques that can be used for risk assessment, but two methodologies for risk analysis are highlighted, which are the quantitative and qualitative method.

Quantitative Method

Such as determining the value of the loss expressed as a percentage of the value of the goods or in absolute values or the size of the loss to be incurred in the future.

According to De Castro et al (2005:24) quantitative risk analysis depends on obtaining and weighting two parameters: the frequency or probability of a given phenomenon occurring, and the magnitude of the socioeconomic consequences associated with them. Thus, the most generic equation to express risk would be given by: $\mathbf{R = P \times C}$, where $\mathbf{P}$ = probability of occurrence of the process in question, and $\mathbf{C}$ = social and economic consequences.

Qualitative Method

Like the identification of the type of risk (it is a risk of fire, explosion, electrical damage, etc.) The association of qualification and quantification gives us the idea of the size of the risk. See tables 1, 2, 3 and 4

Table 2: Quantitative risk analysis in terms of intensity

Level of Disaster	Features
Level I Despicable	Small losses are more easily bearable and surmountable for the affected communities.
Level II Marginal	The damage caused is of some importance and the damage, although not large, is significant.
Level III Criticizes	The damage caused is important and the damage massive, the situation of normality can be restored but with the source of state resources
Level IV Catastrophic	They are not surmountable and bearable by the communities, the restoration of normality depends on the mobilization and coordinated action of the three levels of the civil defense system and in some cases international aid.

Source: Adapted from SAITO, Silva M. (2010:5).

Table 3: Qualitative risk analysis of the evolution of Disasters

Classification	Examples
Sudden or Suddenly Advancing Disasters	Earthquakes and floods

Gradual or chronically evolving disasters	Stretching
Disasters by addition of partial events	Traffic accidents

Source: Adapted from SAITO, Silva M. (2010:5).

Table 4: Analysis of quantitative levels of risk impairment

Level of Damage	Share of loss in GDP
IV	Loss> 30% GDP
III	10% <Loss ≤ 30% GDP
II	5%<Damage ≤ 10% GDP
I	Damage ≤ 5% GDP

Source: Adapted from SAITO, Silva M. (2010:5).

Table 5: Quantitative levels of risk acceptability

Risk level	Name	Description
1	Despicable	Acceptable
2	Minor	Acceptable subject to improvement
3	Moderated	Acceptable sporadically
4	Critic	Not acceptable
5	Catastrophic	Absolutely not acceptable

Source: Adapted from SAITO, Silva M. (2010:5

With all the qualitative and quantitative techniques, they involve calculations although the qualitative one uses simpler calculations which provide subjective data while the quantitative one presents results based on objective values (Navarro, 2004:2).

1.2 Vulnerability Calculations

According to Cutter (2011:2) vulnerability in a broad definition *"is the potential for loss"*, includes both elements of exposure to laughter (the circumstances that put people and localities at risk in the face of a particular hazard) and of propensity (the circumstances that increase or reduce the capacity of the population, infrastructure or physical systems to respond to recovery from environmental threats), and the concept of vulnerability as a global dimension according to Dutra et al, (2014:400) incorporates risk theory in view of the different dimensions closely linked to each other. The canonical expression of the definition of quantitative risk is currently presented in several ways and one of the most effective in risk analysis and management according to De Almeida, (2005:21) is the following:

Eq.1: Risk Calculation

$$\boxed{\text{Risk} = P \times E \times V}$$

Source: De Almeida (2005:21)

Where:

P- Hazards or probability of the process occurring with magnitude M (destructive potential).

E - Exposure "values" of goods exposed to potential hazardous impacts with a certain magnitude M.

V - Vulnerability (physical) degree of damage or loss of exposure value as a result of impact. It is noted that the distinction between exposure (still undamaged or potentially vulnerable) and physical vulnerability (harm operator in action) as a function of the intensity of impact in each scenario is highly effective operationally and provides a basis for framing mitigation measures, however;

Eq.2: Calculation of Vulnerability

$$\boxed{\text{Vulnerability} = \text{To the value of the damage / Value of the Expose}}$$

Source: De Almeida (2005:21)

1.3- Vulnerability Classification

The study of the different vulnerabilities depends from the very beginning on the types of risks considered since different dangerous processes affect different elements in different ways, also provoking different reactions in the search to resist them or to recover them (CUNHA, 2006:159).

Therefore, vulnerability to risks is classified according to the model below:

Figure 9: Environmental Risk Classification Model

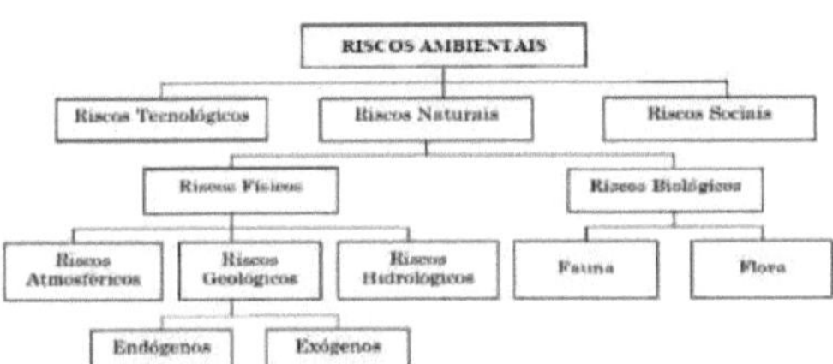

Source: Cerri & Amaral (1998: 302)

Taking this classification as a reference, the object dealt with in this work is the environmental risk in the scope of vulnerability to hydrological risks with a focus on water scarcity.

1.4 Vulnerability Analysis

According to Alcântara (2012:37) in the vulnerability analysis two variables must be considered:

- ✓ Exhibition
- ✓ Disability

In the exposure are considered the elements that are in the area of risk, that is, susceptible to some damage such as: population, social networks, goods and services, soil degradation, deforestation, climate change, among others and the disability concerns everything that does not allow the individual, group or community to deal with the natural hazard, especially groups with less economic conditions and lack of information.

Second Green: (2008:15) *"the degree of loss to which a particular element is subjected is expressed on a scale ranging from zero (0), no damage occurs, to one (1) which means that the damage is total resulting in the destruction of the element at risk".*

1.5 Vulnerability Analysis as Damage Exposure

According to OECD (2012:31) The factors, processes and conditions that drive vulnerability should also be identified and analysed together with the nature and extent of vulnerability that can be classified through various dimensions:

✓ Physical Dimension

The physical dimension concerns the quality and robustness of buildings and other infrastructure as well as the quality and efficiency of infrastructure prevention.

✓ Human and Social Dimension

It is the social and health fabric of the population including physical health, level of education and education of the population, peace and security, equity and social solidarity.

✓ Economic and Financial Dimension

It corresponds to the economic fabric as well as financial resources, wage equity, productivity and level of financial protection.

✓ Environmental Dimension

It is the quality and diversification of environmental resources such as biodiversity, water, soil, air and the availability of natural resource services such as access to clean air, access to drinking water, access to food.

✓ Institutional Dimension

It is the quality of institutional governance that limits itself to making the right decisions, but weakness in these dimensions contributes as qualitative factors of vulnerability.

Figure 10: Model of the Water Scarcity Vulnerability Component Representation

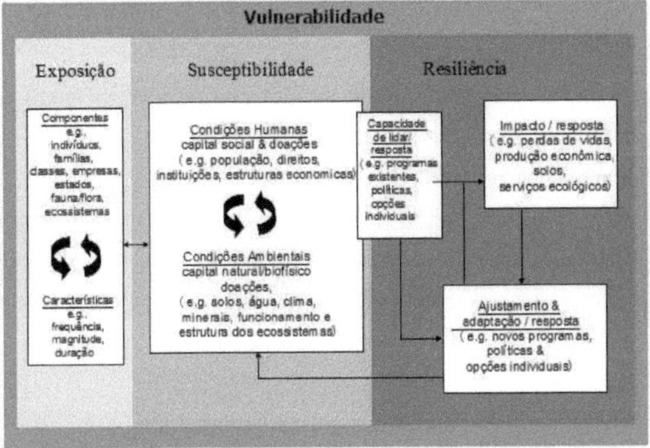

Source: Amaral, (2010:199)

1.6 Social Vulnerability

According to Mendes et al, apud Kuhlicke et al (2011:96) *"social vulnerability is a product of certain spatial, socioeconomic, demographic, cultural and institutional contexts, so its approach is sensitive to local conditions and the temporal dimension"*. The results of the analysis are placed in a broader context, not only with regard to vulnerability but also to the resistance of populations and to the parallels of risk research.

1.6 Environmental Vulnerability

When conducting environmental studies it is understood that each class of the variable has a weight that influences the increase or decrease of environmental vulnerability in a given area. Standardization allows the values assigned to map classes to be uniform. Thus, the scale of vulnerability of this survey will be presented in the range of values from 0 to 1. The closer to 1 (one) the more vulnerable and the closer to 0 (zero) the less vulnerable. The scale below demonstrates the environmental vulnerability (Rezende apud Paula 2016:63).

Figure 11: Environmental Vulnerability Scheme

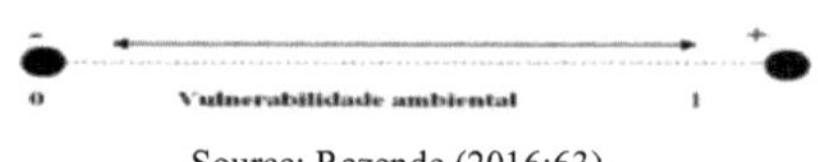

Source: Rezende (2016:63)

1.7 Water Resources

Water should not be confused as a water resource, so Gama, apud Pompeu (2009:37) points out that water "*is a natural element, uncompromised with any use or use but its use can make it a resource.* This point of view is also shared by Junior (2004:3) when defining water resources as "*the portion of fresh water accessible to mankind at the present technological stage and at a cost that can be compacted with its various uses*". The contradiction of water as an element of nature and as a water resource derives from the view that most available water has high salinity levels compared to fresh water (Junior, 2004:3).

Fresh water according to Rebouças et al (2002:3) is designated according to the worldwide criterion for the environmental classification of waters, as that with a salinity content of less than 1000 mg/l. These waters occur in the proportions of emersed land, continents, islands, flowing through rivers, streams, streams, underground deposits, filling the lagoons, dams, forming swamps or pools, and are therefore called inland waters.

In the case of Funhalouro district, the water occurs in underground deposits and to access it is a challenge as the depth of the water table is between 50-75 meters and rainfall is scarce to recharge the aquifers and increase the amount of water in the soil and on the other hand the existing water has a higher content of salinization, which puts at risk the availability of fresh water (PRONASAR, 2011).

1.8 Hydrological Principles of Water Resources

The first great hydrological characteristic of water resources is the totally natural circulation of water caused by solar energy that starts causing the evaporation of lakes, seas, rivers, etc.

According to Kobyama, (2008:47) *"water moves in nature in three phases (gaseous, liquid and solid)"* and this circulation is called hydrological cycle whose components are called processes, of which the following stand out: condensation, precipitation, interception, infiltration, detention, percolation, runoff, underground runoff, river runoff and evapotranspiration (evaporation + perspiration) see figure no. 12

Figure 12: Schematic of the Hydrological Cycle

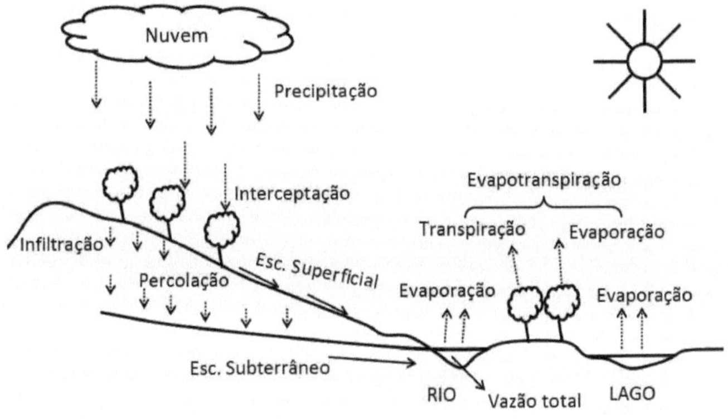

Source: KOBYAMA, Mosato et al (2008:47)

1.9 Hydrological Cycle Processes

According to ANA (2015: 19-21) Processes in the hydrological cycle are characterised as follows:

1.9.1 Precipitation

It is all the water coming from the atmospheric environment that reaches the earth's surface and may present itself in the form of mist, rain, hail, hail, dew, frost, snow and it divides the frontal or cyclonic precipitation, orographic and convective.

1.9.2 Interception

It is the retention of water from part of the precipitation above the surface of the soil, and may occur by the presence of vegetation or other form of obstruction to the flow. The volume is retained by evaporation, returning to the atmosphere. This process interferes with the water balance of the river basin. Interception can occur in two ways:
✓ Plant interception
✓ Depression interception

1.9.3 Plant interception

This type of interception may depend on some variables, among them:
➤ The precipitation characteristic
➤ Weather conditions,
➤ Type and density of vegetation
➤ Period of the year.

The main characteristics of precipitation are intensity, precipitated volume, and previous rainfall. The leaves usually intercept most of the rainfall but the trunk layout contributes significantly.

The time of year can characterize some types of crops that present the different stages of growth and harvest.

The intercepted precipitation calculates as follows

Eq.3: Intercepted Precipitation

$$S_i = P - T - C$$

Source: ANA (2015: 19)

Where:

Si = Intercepted precipitation

P = Precipitation

T = The precipitation that crosses the vegetation

C = Parcela that flows through the trunk of the tree

1.9.4 Storage in Depressions

In the river basin there are natural or artificial obstructions to the flow accumulating part of the precipitated volume. In rural areas, for example, after a flood when areas without drainage form small lakes. The volume of water retained in these areas only decreases by evaporation and infiltration. The volume of water retained by depressions is calculated as follows

Eq.4: Volume of water retained

$$V_d = S_d \left(1 - e^{-kPe} \right)$$

Source: ANA (2015: 19)

Where:
Vd = Retained Volume
Sd = Maximum Capacity
Pe = Effective precipitation
K = coefficient equivalent to 1/Sd

1.9.5 Infiltration

Infiltration is the phenomenon of water penetrating the soil layers near the surface of the ground, moving down through voids under the action of gravity until it reaches a supporting layer, which then retains it forming underground water. The infiltration can be divided into three phases

- ✓ Exchange - the water is near the surface of the ground subject to return to the atmosphere by a capillary aspiration caused by the action of evaporation or absorbed by the roots of the plants and then perspired by the plant
- ✓ Descent - when the movement of water reaches a layer of waterproof soil
- ✓ Circulation - when the accumulation of water occurs where the underground leftovers are constituted.

1.9.6 Evaporation

It is the set of phenomena of a physical nature that transforms liquid or solid water into water vapour from the surface of the soil and transferred in this state to the atmosphere. This process can only occur naturally if there is energy entering the system from the sun, the atmosphere or both and will be controlled by the rate of energy in the form of water vapor that propagates on the surface of the earth.

Evaporation can occur in water bodies, lakes, accumulation reservoirs, straight water in the surface layer of the soil and seas and is influenced by the temperature and relative humidity of the air, wind and steam pressure.

1.9.7 Surface Flow

Surface runoff covers the excess rainfall that occurs immediately after heavy rainfall and moves freely across the land surface (Carvalho & Silva, 2006: 95). Runoff is influenced by climatic factors (related to rainfall) and physiogeographic factors (related to the physical characteristics of the basins).

1.9.8 Coefficient of Surface Flow (C)

It is the ratio between the volume of water drained superficially and the volume of water precipitated. This coefficient can be relative to isolated rainfall or relative to a time interval where several rains occurred and calculates itself as follows:

Eq.5: Coefficient of Surface Flow

$$C = \frac{\text{Volume total escoado}}{\text{Volume total precipitado}}$$

Source: ANA (2015: 21)

Where:

C - Surface Flow Coefficient

1.10 Freshwater Distribution Worldwide

According to Tundisi apud Viriato et al (2015:2) 2.5 % is fresh water, 68.9% is glaciers and polar ice caps in mountainous regions, 29.9% is groundwater, 0.9% is the moisture content of soils and marshes and only 0.3% is the fresh water stored in rivers and lakes, actually available for use in different activities.

According to Cunha et al, (2009:4) The distribution of fresh water worldwide is as follows:

Figure 13:World Water Distribution Chart

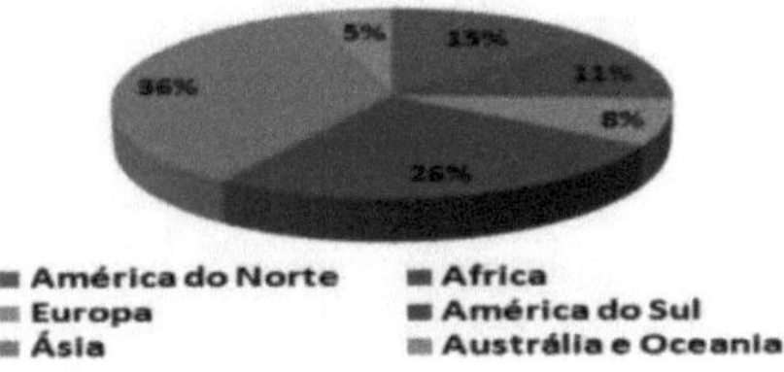

Source: CUNHA et al, (2009:4)

This figure shows that water resources are unevenly distributed around the world and that in some regions they are objects of authentic wars.

1.11 Water Resources Uses

According to Cunha et al, (2009:3) The use of fresh water differs between developed and developing countries.

1.11.1 Fresh water consumption in developed countries

- 30% In agriculture
- 11% In domestic consumption
- 59% In industry

1.11.2 Developing Countries

82% in agriculture;

8% In domestic use;

10% in industry.

As previously mentioned, water is a very abundant element of nature that for many years it was thought that the lack of drinking water was impossible, which caused no concern for the preservation of this element as a resource and today the diverse and numerous uses of water are contributing to its scarcity and contamination (MOTA et al, 2008: 9)

1.12 The Difference Between Water Scarcity and Drought

The concept of drought is closely related to the observer's point of view although the primary cause of droughts is insufficient or irregular rainfall, there is a sequence of causes and effects (Campos, 1994:14).

41

In this same thought, Viva & Maia (2010:30) defers drought, "as *a natural phenomenon that can assume extreme consequences, as a transitory anomaly of the transition conditions of rainfall in a given area during a certain period of time and scarcity as a phenomenon corresponding to an excess of demand over existing natural assets reflecting a long-term assessment"*. Another perception summarizes the difference of these concepts by stating that drought is a natural phenomenon caused by the absence of rainfall while scarcity is a social and political problem when consumption exceeds the physical availability of water (Cunha et al, 2009:15). In the absence of a universally accepted definition, three essential components are pointed out to determine water scarcity:

- ✓ The existing needs
- ✓ The fraction of water that can be mobilized
- ✓ The long-term time and space scales

However, it is necessary to stress that a drought situation can then enhance and or aggravate situations of imbalance between natural availability and needs for the main uses in any hydrographic region (Viva & Maia , 2010:30).

From this point of view, looking at the needs component in the Administrative Post of Funhalouro according to (ESF,2011:49) there is divergence between the criteria used by DNA and INE, because for DNA Funhalouro appears as a district in which more than half of the 23,734 known inhabitants have access to water, since 57% of the communities have an operational source and sufficient for their inhabitants, but for INE, the previous results are compromised since 77% of the known population is below the percentage of minimum coverage (100%). This implies that in 66 communities of the 99 in the Administrative Post of Funhalouro there is no access to water. With everything done by Engyneria Sense Fronteres in 2011 found that the criteria used by INE is more real in Funhalouro district than the criteria used by DNA.

1.13 Causes and Consequences of Water Scarcity

There is great controversy over the causes of water scarcity, with experts pointing to the 21st century water crisis as a management problem rather than a real crisis of scarcity and stress. Others point to it as a result of a set of environmental problems aggravated with other problems related to the economy, see the figure below.

Figure 14: Flowchart of Water Shortage Environmental Processes

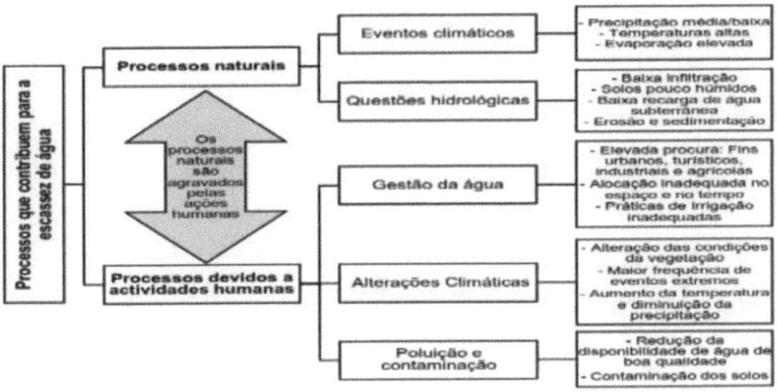

Source: PERREIRA, (2009:12)

and for others is the result of the worsening and complexity of the water crisis resulting from real problems of availability and increasing demand and from a still sectoral management process and response to crisis and problems without predictive attitudes and systematic approach. (TUNDISI apud Roger et al, 2008:7).

According to Pereira et al (2002:5), the natural or environmental causes of water scarcity based on aridity, drought and desertification are defined as follows:

1.13.1 Aridity

It is a natural, permanent climatic feature, corresponding to an imbalance in water availability, with low average annual precipitation, with high spatial and temporal variability, resulting in a climate of low overall humidity where ecosystems have low carrying capacity.

1.13.2 Drought

It is a temporary imbalance of water availability consisting of relatively lower than average rainfall. It presents an uncertain, unpredictable or difficult to predict frequency, duration and severity, resulting in diminished water resources and reduced carrying capacity of ecosystems.

According to Cunha et al, (2009:3) according to the places and climates the drought can be studied as follows:

Table 1: Types of drought

Type of drought	Feature
Permanent drought	Existing in desert climates, it is characterized by the total absence of water. Agriculture is totally impossible without the aid of artificial and permanent irrigation. The existing vegetation is totally adapted to the lack of water
Seasonal drought	It is a particularity of regions where the climate is semi-arid. The humidity is low as is the precipitation. Agriculture is only possible in rainy times
Irregular and variable drought	Irregular drought can occur in any region where the climate is humid or sub-humid and is characterized by being dry whose return period is brief and usually limited in areas rather than large regions. It does not occur in a defined season and its occurrence is unpredictable.

Source: Adapted based on MICOA data (2005:5).

The southern region of Save mainly Gaza, Inhambane and northern Maputo provinces are considered to be at high risk for rainfed agriculture as the average annual rainfall does not exceed 400mm, insufficient for example to meet the water needs during the vegetative cycle of maize which is the food base of the population. In general terms, climate change, deforestation, uncontrolled burning, overgrazing, soil erosion and inappropriate agricultural practices are cited as causes of drought in these regions (MICOA, 2005:28).

1.13.3 Desertification

Desertification consists of the permanent imbalance in the availability of water, induced by man through inadequate use of land and water, ineffective measures relating to the maintenance of territory and human-induced climate variation on a global scale

1.14 Social Context of Water Scarcity

For the social, economic context Tundisi (2008:9) highlights the following as processes of the water crisis:

- ✓ Intense urbanization and demand for water
- ✓ Increased vulnerability of the human population due to contamination and difficulty of access to good quality water (drinking and treated)
- ✓ Problems in the lack of articulation and consistent actions in water resource governance and environmental sustainability

1.15 Consequences of Water Scarcity

According to Morrison (2009:5) and Cavalcante, (2008:126) water scarcity is the consequence:

- ✓ High cost of water
- ✓ Constant complaints about the use of water
- ✓ Conflicts between local communities and larger scale users
- ✓ Difficulties in the development of agriculture thus generating hunger,

In addition to the causes and consequences already mentioned, the following can also be mentioned:

Table 2: Causes and consequences of Water Scarcity

Atividade humana	Impacto nos ecossistemas aquáticos	Valores/serviços em risco
Construção de represas	Alteração do fluxo dos rios, transporte de nutrientes e sedimentos, intereferência na migração e reprodução de peixes	Habitats, pesca comercial e esportiva, deltas e suas economias
Construção de diques e canais	Destruição da conexão do rio com as áreas inundáveis	Fertilidade natural das várzeas e controles das enchentes
Alteração do canal natural dos rios	Danos ecológicos dos rios. Modificação dos fluxos dos rios	Habitats, pesca comercial e esportiva. Produção de hidroeletricidade e transporte.
Drenagem de áreas alagadas	Eliminação de um componente fundamental dos ecossistemas aquáticos	Biodiversidade. Funções naturais de filtragem e reciclagem de nutrientes. Habitats para peixes e aves aquáticas.
Desmatamento/uso do solo	Mudança de padrões de drenagem, inibição da recarga natural dos aquíferos, aumento da sedimentação	Qualidade e quantidade da água, pesca comercial, biodiversidade e controle de enchentes.
Poluição não controlada	Prejuizo da qualidade da água	Suprimento de água. Custos de tratamento. Pesca comercial. Biodiversidade. Saúde humana.
Remoção excessiva de biomassa	Diminuição dos recursos vivos e da biodiversidade	Pesca comercial e esportiva. Ciclos naturais dos organismos.

Source: DA PAZ (2004:12)

1.16 Societal Environmental Value of Water

In the environmental field, the distribution of water has an influence on vegetation as well as on the faunal species found in a certain region and plays an important role in the shape of the landscape. Any change in water regime will inevitably change all local environmental components. With regard to the social dimension water is essential for life and affects lifestyles e.g. in areas of water scarcity agriculture and livestock influence considerably the use of water and there is a need to capture rainwater for domestic use to retain a certain amount possible in the soil for crop growth, water falling from the roof or compacted soil can still be collected for storage. On the other hand, the lack of hygiene or poor sanitary practice is a factor in several diseases such as diarrhoea, cholera, typhoid, hepatitis which result from the transfer of bacteria by dirty hands or dirty water to the mouth and intestine, thus developing an enteric infection (Perreira et al, 2002:35)

The main sectors that use water in Mozambique are irrigated agriculture, domestic and industrial use according to the figure below.

Figure 15: Chart of Distribution of Water Use Sectors in Mozambique.

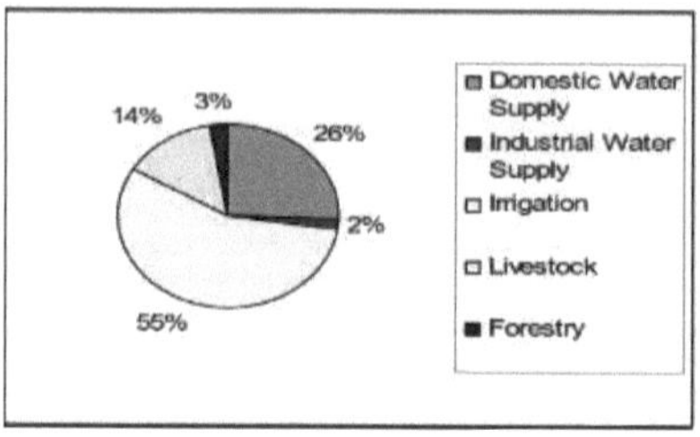

Source: World Bank (2007:15)

1.17 The Global Water Scarcity Situation

According to Jacobi, (2017:54 apud WWDR) and UNESCO (2015) *"one billion people currently suffer from water shortages and around 40% of the world's population live in countries under water stress",* see figure 16. Five of the ten most densely populated river basins on the planet, such as those of the Yang-Tsé River in China and the Ganges River in India, are already exploited above levels considered sustainable. Africa, which has the highest prevalence rate of hunger, is also the second driest inhabited continent in the world, behind Oceania. In the last 30 years, 57 million people have been affected by drought in Ethiopia. In India, more than 70 percent of the rains occur in just three months of the year, which causes water shortages for much of the year in non-irrigated agriculture. Water scarcity in the world is aggravated by social inequality and a lack of sustainable management and use of natural resources.

According to the figures presented by the UN, it is clear that controlling the use of water means holding power.

Figure 16: Map of areas with greatest threat to water security and biodiversity

Source: Vörösmarty, C. J., et al. (2010).

In the summer of 2011, in Somalia, Kenya and Ethiopia more than 10 million people were affected by the worst drought in 50 years, tens of thousands died of hunger (Estarques,2013:2).

1.17.1 Australia

In early 2009 Australia experienced a food crisis due to drought that lasted roughly 7 consecutive years directly affecting the lives of thousands of Australian farmers, where the prolonged absence of rainfall led to impoverishment, bankruptcy, as well as profound changes in living habits. In a region that once exported cereals all over the world, a reverse situation is now prevailing, depending on the export of cereals from other countries. (Serra, Apud, National Geograph, 2012:57)

1.17.2 Kenya

In 2008 about 4 million Kenyans were dependent on food aid, most of which they could not plant anything until the first half of 2009 (Serra, Apud PMA, 2012:57).
In the second half of September 2009, the press reported violent clashes in northern Kenya, motivated by the dispute over fertile land and access to water following one of the worst droughts in living memory.

1.17.3 Sahel Region

Throughout 1972 - 1975 and 1984 - 1985 it is estimated that more than 1,000,000 (one million) people died from drought and famine. (Henson et al, 2009:56).

1.17.4 Tanzania

In the years 1964 to 1965 Tanzania lost a significant crop production of about 3% of $1.1 billion. (Burton et al, 1993:69).

1.17.5 Mozambique

In 1993, data on drought indicate that in the district of Funhalouro about 50% of the population affected by drought died (MICOA,2005:36).

According to estimates by the International Food Policy Research Institute, by 2050 a total of 4.8 billion people will be under water stress. In addition to problems for human consumption, this scenario, if confirmed, will put agricultural crops and industrial production in jeopardy, since water and economic growth go hand in hand (Jacobi, 2017:54).

1.18 Water Scarcity Adaptation Strategies in Some Countries of the World

According to Martins, (1983:101) the concept of strategy derives from the term "strategists" which "*means the action of commanding armies.* "For Ferreira (1986:33) the concept of strategy is defined as "*the science and art of using available knowledge and personal and technical means to achieve the timely solution of the problems to be faced*" while adaptation is the change of behaviour and activities in response to climate change, not only to protect them from negative impacts but also to benefit from any positive effects that may occur (MAMAOT, 2013:4).

Nations in arid areas have developed appropriate skills to use and manage water in a sustainable way. These include measures to avoid water wastage, the adaptation of appropriate technologies to prevailing conditions by making successful use of water for agriculture and other productive activities, and the development of institutional activities and regulatory conditions that widely influence the behavior of rural and urban societies (Pereira, 2002:5).

The drought according to Perreira (2002:5) requires appropriate approaches as there is a need to understand the characteristics and consequences of those phenomena that make water scarcity due to drought very different from the scarcity caused by aridity. Dealing with water scarcity resulting from aridity generally requires engineering and management measures that produce conservation and perhaps seasonal increase of the available resource. On the other hand droughts require the development and implementation of preparedness and emergency measures.

In this context, some strategies implemented by some countries are presented below:

1.18.1 Brazil

According to Nobrega et al (2016:2) in their study on the feasibility of using cisterns in the semi-arid region of Brazil, they concluded that the construction of cisterns is a viable alternative for places where droughts occur and that the capture occurs only by rainwater that falls on the roof of their houses or is supplied by a "Pipa" truck.

1.18.2 South Africa

According to Sigenu (2006:85) in his study the role of rural women in mitigating water scarcity found that one of the strategies found by the Ndonga women's community is the sharing of farmland where they are allocated to a single area of agricultural implements needs, from land preparation, sowing and harvesting which is then divided among the participants. This strategy aims to reduce the irrigation of extensive agricultural land, thus saving water

1.18.3 Namibia

According to Scott (2018:10) Namibia is considered the country most devastated by drought in the southern region of the Sahara desert, compromising the supply of water to the population of Windhoek, led the discussion to two potential solutions to the drought. The first solution has to do with the technical issue and the other has to do with political issues for water governance.

In the technical question it was suggested:

- Rehabilitation of existing water infrastructures such as dams.
- The transfer of the waters of the River Kunene in the north of the Can basin,
- The desalination of water

The issue of water governance was more focused on activating water management

1.18.4 Kenya

The causes of water scarcity are drought, forest degradation, poor water supply management, contamination of well water by latrine microorganisms, population growth (Marshall, 2011:43). The strategy used to face this scenario is:

- ✓ Promote the sustainability of environmental management
- ✓ The management of information on water resources
- ✓ The collection and storage of water through dams

1.18.5 Australia

According to Chartres & Williams (2006: 21), *"climate change, changes in vegetation cover, demographic increase"* are cited as causes of water shortage. The strategies used to address this situation are:

✓ The implementation of the water industry to treat wastewater and the desalination of water,

✓ Detect water leakage in supply systems

✓ The use of solar energy for water desalination

✓ Use of remote sensing technology to improve the perception of the vertical and horizontal distribution of drinking water and saline water resources

1.18.6 Mozambique

In Mozambique, according to the World Bank (2007:7) *"the country is vulnerable to water scarcity due to dependence on international river basins and climate variability (annual and inter-annual variations of precipitation, drought and frequent floods)"*. This situation is also aggravated by underdeveloped and largely degraded water infrastructure, which increases water vulnerability and presents a serious risk to the national economy. As a result, while the population of Mozambique is expected to grow at an annual rate of 2.2-2.3% by 2020, about 70% of the rural population does not have access to a source of water supply (idem, 2007: 12).

According to MICOA (2013:17), *"most actions to ensure the availability of water in quantity and quality also involve the construction and rehabilitation of infrastructure for water management such as natural dykes, dams, mini-hydro channels, community dams, rainwater collection systems (gutters and cisterns)"*. Besides this dimension of creating water management capacity, the strategy also aims at sanitation, aiming at the entire cycle of water use from its collection, through its adequate treatment to ensure quality according to use and also its subsequent deposition in the natural environment in patterns that do not compromise the receiving environment.

1.19 Risk and Vulnerability Situation of Mozambique in Water Scarce

According to ORR, (2009:8), the risk of water scarcity can be understood in terms of insufficient water to meet basic needs as well as the consequences that arise from the situation such as political and economic instability and the risks that arise from unfavourable policies in the search for responses to the water crisis.

Mozambique is historically the country most affected by natural disasters in southern Africa and drought is one of the frequent phenomena whose impact on the lives of populations has been greater than floods, as it causes long-term economic disruption. Although drought vulnerability is high in the southwest (west of Gaza province) and central regions (west of Tete province), small- and large-scale irrigation systems have also helped to reduce the impact of regular droughts.

According to MICOA (2005:3) the vulnerability of these regions is partly due to irregular and unpredictable rainfall as well as the uncertainty of the rainy season which often does not start according to forecasts resulting in erratic sowing periods. These phenomena cause the lowering and salinization of the groundwater table, decreased availability of water underground with strong implications on the annual average evapotranspiration.

The visibility of natural disaster management in Mozambique is still very low despite the fact that the country is extremely vulnerable as institutions such as the INGC responsible for coordinating disaster management are marked by the reactive character of their interventions, where the emphasis is on emergency operations and humanitarian aid and less on risk management, understood as the combination of the threat of an extreme event and specific conditions of vulnerability (MICOA, 2005:29)

1.20 International Water Conventions and Protocols

There are several international water conventions such as the Luso-Spain Convention, the Barcelona Convention, the Rhine Convention, the Convention on the Protection of the Danube River, the Espoo Convention and the Aarhus Convention, but in this work I will highlight the following conventions and protocols:

1.20.1 SADC Protocol on Shared Water Courses

Figure 17: Map of shared watercourses in SADC

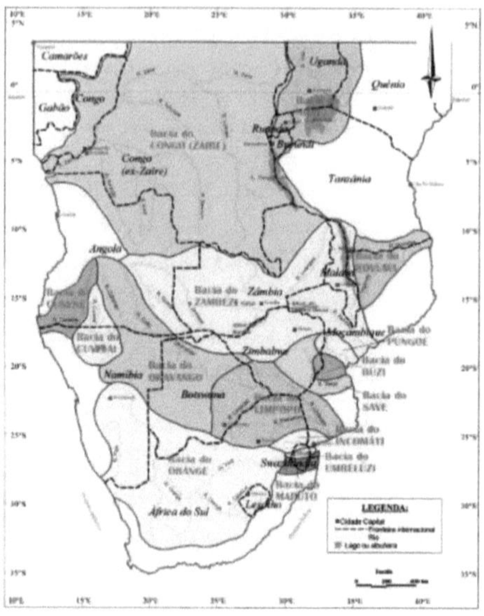

Source: DNA, 1999

The SADC protocol was ratified in 2003 by the member countries:

- Develop a policy for the monitoring of shared watercourse systems
- Promoting equitable use of shared watercourse systems
- Formulate strategies for the development of shared watercourse systems
- Monitor the implementation of integrated water resources development plans in shared watercourse systems.

Source: www.kunene.riverawarenesskit.com- consulted on 03 August 2018

1.20.2 United Nations Conventions

According to Correia & Henriques (2010:13) The following are part of the united nations conventions:

✓ The protocol convention on water and health

- ✓ The Convention of the Protocol on Civil Liability for Transboundary Damage Caused by Hazardous Activities
- ✓ Convention on the Protection and Use of Transboundary Watercourses and International Lakes.

This convention entered into force in 1996 and was ratified by 36 countries of the European Community and the UNECE:

1. Strengthen local, national and regional measures to protect and ensure the ecologically sustainable use of transboundary surface and groundwater;
2. Prevent, control and reduce emissions of hazardous waste from acidification and eutrophication of substances in the aquatic environment
3. Conservation and protection of ecosystems

1.20.3 Water Scarcity Conventions (The Hague Declaration)

The Hague Declaration according to Correia & Henriques (2010:13) was signed on 22 March 2000 in The Hague, Netherlands. This declaration was intended:

- Ensuring water security in the 21st century which means ensuring that freshwater reserves and cohesive ecosystems are protected and restored; that all people will have access to safe and sufficient water at a cost compatible with a healthy and productive life

1.20.4 Declaration on Water and Sustainable Development

Adopted in Dublin (1992) by the International Conference on Water and the Environment. The declaration affirms the need to value and optimise the use of water resources in the following principles

- ✓ Fresh water is a finite and vulnerable reuse, essential for the maintenance of life, development and the environment
- ✓ Water development and management should be based on a participatory approach, involving usurers, planners and politicians at all levels
- ✓ Women have a central role in the provision, management and preservation of water
- ✓ Water has an economic value in all its multiple uses and should be recognised as an economic asset.

1.20.5 Interim Tripartite Agreement between the Republic of Mozambique, the Republic of South Africa and the Kingdom of Swaziland (Resolution n° 53/2004 of 1 December 2004)

According to the Bulletin of the Republic, (2004) It was signed on 7 August 2000 and aims at cooperation on the protection and use of water resources of the Incomati and Maputo watercourses.

1.21 Sustainable Development Goals (SDOs)

According to UNDP (2015:7) on water, set as target 6 to ensure the availability and sustainable management of water and sanitation for all, however, these targets have as targets:

> By 2030, achieve universal and equitable access to safe drinking water accessible to all

> By 2030, substantially increase the efficiency of water use in all sectors and ensure sustainable withdrawals and fresh water supplies to address water scarcity, and substantially reduce the number of people suffering from water scarcity

> By 2020, protect and restore water-related ecosystems, including mountains, forests, wetlands, rivers, aquifers and lakes

In 1991 the Water Law was passed, which emphasises the fundamental principles of state action in the water sector, aiming at the continuous and sufficient supply of drinking water to meet domestic and hygienic needs. As a follow-up to this regulatory effort, the National Water Policy was approved in 2005, which considers water as an asset of social value (although it also recognises its economic value). This national water policy has as its objectives:

a) Integrated water resources management - Water resources will be managed in an integrated way based on the river basin as a fundamental and indivisible unit. Management and planning must respect the intrinsic link between surface water groundwater, quantity and quality aspects of water from source to mouth, environmental conservation and development needs.

b) Meeting the basic needs of the poorest population - The government gives high priority to meeting the basic needs of the poorest rural and urban population in terms of adequate water supply and sanitation, always seeking a situation of sustainability, with the ecective participation of the beneficiaries in defining the solutions to be adopted

c) The economic value of water - Water is important for economic development and poverty reduction. To enable services to become financially viable, the price of water should approach its economic value.

1.22 National Water Resources Management Strategy

According to the Council of Ministers (13:2007), Mozambique's challenges in managing and developing water resources to meet the goals of the action plan for the reduction of absolute poverty and the Millennium Development Goals include drinking water and sanitation, ecosystem conservation, disaster mitigation and risk management. All due to the current pressure on water resources and scarcity lead to the adoption of strict measures to improve water use efficiency in order to make water more available to meet different demands and to this end, the strategic policy plan for the water sector in Mozambique as illustrated in figure 18.

Figure 18: Strategic Organigram - Water Sector Policy in Mozambique

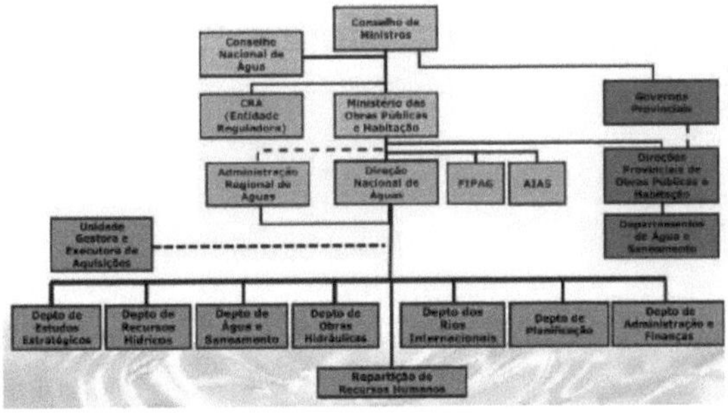

Source: Ministry of Public Works, 2012

1.23 Water Supply in Rural Areas

The policy guideline contemplates the expansion of water supply to rural areas and reduction of regional asymmetries; promotion of the demand principle in the water sector, with the education, health and agriculture sectors as a basis for ensuring community involvement and creating a sense of ownership to ensure the sustainability of rural systems. Participation of women, stakeholders and the private sector should be promoted

Figure 19: Graph Water Supply in Rural Areas

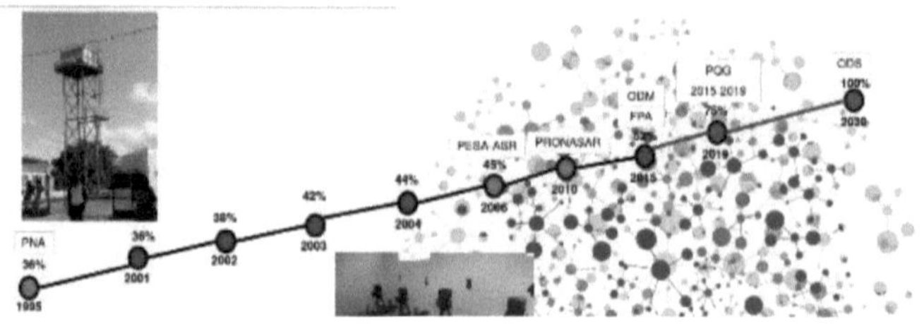

Source: Correia, (2017: 12)

1.24 Sanitation

The policy includes expanding sanitation services by strengthening institutional arrangements for operation and maintenance; applying an integrated water supply and sanitation policy; strengthening the role of local authorities; implementing a financial policy in which the Government promotes investments necessary for the rehabilitation and expansion of infrastructure; creating mechanisms and incentives to promote service provision in peri-urban areas, promoting the role of women, stakeholders and the private sector and promoting hygiene.

With the implementation of the National Water Management strategy for rural areas, the number of people with access to water is growing, which could lead to the achievement of sustainable goals in 2030 as illustrated in the figure below.

1.25 Water Resources and Climate Change in Mozambique

Currently, it circulates in the vast literature that discusses the subject of climate change, three lines of thought. One composed by researchers who consider global warming as the responsibility of human activities in the process of production and reproduction of geographical space, with greenhouse gas emissions as the main responsible. This group does not rule out the possibility of warming caused by natural factors, but with little significance. A second current of researchers highlights that the planet earth experiences a long cycle of temperature variation, with the warming provided only by natural processes. Another current of researchers defends the human interrelationship and natural phenomena as the main responsible for the current stage of climate change (Dos Santos et al, 2010:64)

Climate changes alter oceanic and atmospheric circulation patterns, modifying hydrologic cycles and precipitation patterns. As a consequence, the water supply on the surface of the earth changes to more (flooding) or less (scarcity) and these changes in the water supply on the surface put at risk the water security of human populations on a smaller or larger scale (Lima,2012:15).

1.26 Relationship of El Nino Phenomenon, South Oscillation (ENSO) to Water Scarcity

According to Weng et al (2016:86) "*el nino is an oceanic atmospheric phenomenon that is characterized by hot and cold episodes associated with changes in normal sea surface temperature (TSM) patterns and trade winds in the equatorial Pacific Ocean that lead to changes in precipitation behavior around the world.* The el Nino phenomenon has two phases: a hot (el nino) and a cold (la nina), both oceanic phenomena and the southern oscillation is of atmospheric nature. As for the atmospheric component, the studies of Sir Gilbert Walker at the beginning of the 20th century showed an inverse corellation between the pressure on the surface over the Pacific and Indian Oceans called the southern oscillation: when the pressure is high in the Pacific Ocean it tends to be low in the Indian Ocean (Da Cunha, 2011:1).

1.27.1 Impact of Enso on the World

During the episodes of the warm phase of Enso, rainfall increases over the central pacific, south central pacific, southwestern South America and eastern equatorial Africa according to the figure n°20. In these episodes the regions of northeast South America, southeast Africa, Indonesia, New Guinea and the western Pacific and Australia, with the exception of the southeast of this continent, suffer drought (Couane apud Francisco, 2015:14).

Figure 20: Illustrative map of ENSO Oscillation

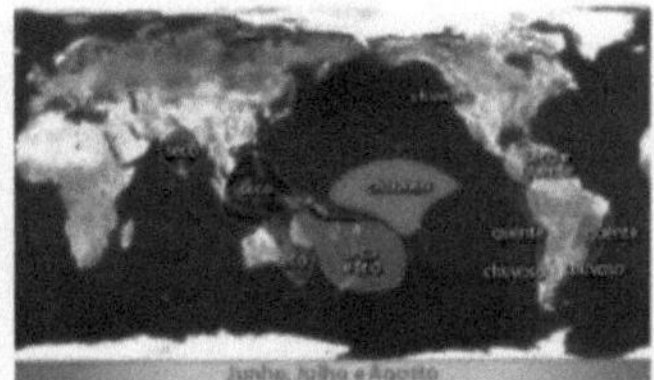

The cold phase of this phenomenon La nina is globally characterized by rainfall above normal conditions west of the equatorial ocean in the Northern Hemisphere and over northern Australia and Indonesia during the winter and the Philippine during the summer of the Northern Hemisphere It is wetter than normal conditions in Southeast Africa and Brazil during the winter. See and figure n°21. On the west coast of tropical South America (Gulf Coast) and South America (south of Brazil to central Argentina) normal conditions are drier. The high rainfall patterns of La Nina are allied to the heat after El Nino and the air in this phase finds s hotter containing more water vapor. Generally La Nina occurs after el Nino (Couane apud Francisco, 2015:14).

Figure 21: Illustrative map of the La Nina phenomenon

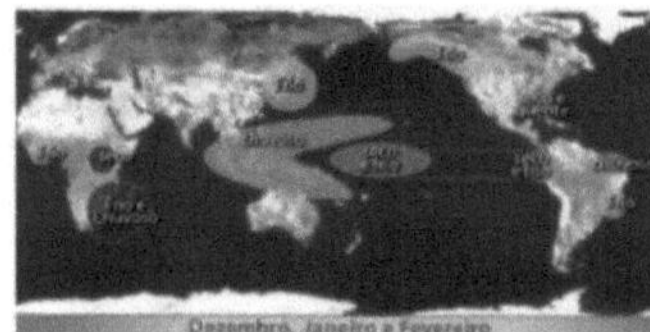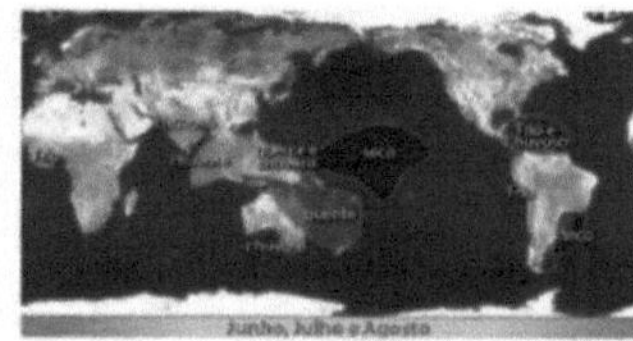

Source: CPTEC (2010)

1.27.2 Impact of Enso on Southern Africa

Southern Africa is predominantly a semi-arid region with unstable rainfall characterized by frequent droughts and floods. It is also widely recognised as one of the most vulnerable regions to climate change due to its low levels of adaptive capacity in rural communities and high dependence on rainfed agriculture (IPCC, 2007).

In southern Africa the hot events of Enso (El Nino) are related to the drought situation, so the negative indices of southern oscillation or the positive anomalous indices of temperature correspond to dry periods while the positive or negative anomalous indices of TSM (La Nina), correspond to wet periods or excess rainfall, Couane apud Benessene, (2015: 16), see figure below.

Figure 22: Map of El Nino and La Nina Evolution in Southern Africa

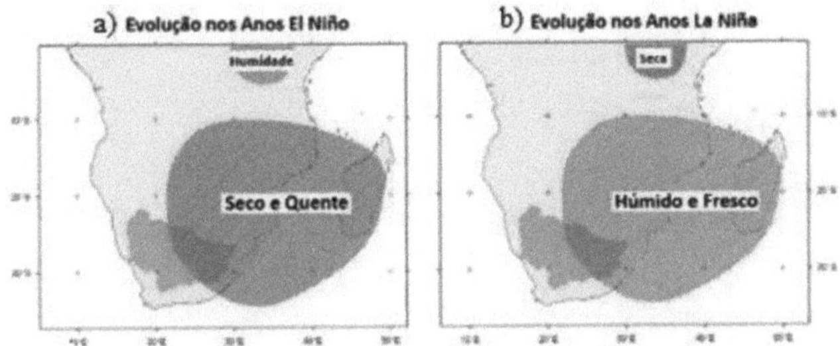

Source: www.limpoporak.com quoted by Couana, (2015: 16)

1.28 Impact of Climate Change on Water Resources

According to Devis (2011:60) *"Climate instability is likely to further compromise water resources by ultimately threatening the sustainability of future availability that depends on supply and demand constraints.* "The greatest likely risks to water resources in the southern African region are:

- ✓ Less water availability in rivers resulting from the net effect of rising temperatures and higher evaporation together with changes in timing and amounts of precipitation
- ✓ Increased risk of water pollution and its poorer quality associated with erosion and high rainfall phenomena

In the case of Funhalouro District as in other districts of Mozambique experience shows that drought and floods have negative impacts in different areas of activity and can cause different effects such as: loss of crops, drying of water points (wells, lagoons, lakes, streams, rivers, etc.), reduction of grazing sand, rising prices of agricultural products, loss of human life, outbreak of diseases and loss of biodiversity. See figure n°18 the districts at risk of drought and flooding in Mozambique (MICOA, 2007:13)

Figure 23: Maps of drought and flood risk areas in Mozambique

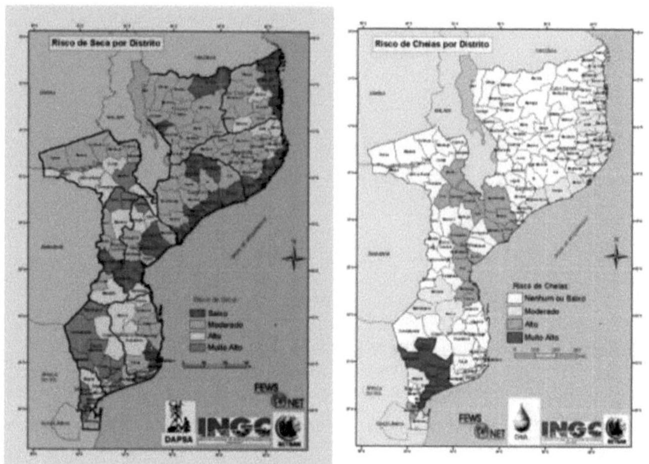

Source: MICOA (2007:15)

A greater focus on the management and prevention of scarcity and droughts could facilitate the process of adaptation to climate change through the flexibility of analysis and simulations of future scenarios in comparison with the development and previous actions described (MAIA, 2010: 25). Within this vision the analysis of past and present impacts according to MICOA Apud INGC (2013:17) *"climate change in Mozambique is manifested mainly in temperature patterns, precipitation"*. These patterns were projected in a simulation of scenarios on the severity of the impacts of climate change until 2025 according to Table 6.

Table 6: Impact of Climate Change

Sector/área	Impactos das MC						
	Mudança nos padrões de temperatura atmosférica	Mudança nos padrões de precipitação	Secas	Cheias	Ciclones tropicais	Subida do nível de água do mar	Aumento da temperatura média do mar
Recursos hídricos	•••	•••	•••	•••	•••	••	•

Chave: ••• Elevado; •• Moderado; • Baixo (ou não conhecido)

Source: MICOA (2013:7)

1.29 Methodologies for Socio-Environmental Vulnerability Analysis at Water Scarcity

The methodological issue for the analysis of socio-environmental vulnerability to water scarcity is widely discussed by the various literatures that address this issue, but according to Rijsberman (2006):2) to assess the security of water needs for human use and environmental services *"an analysis of the specific water quality desired with the quantity of water available or that can be made available within time and space is necessary"*, it is within this context that in the last 20 years many indexes and indicators have been developed to quantitatively assess the vulnerability of water resources, in particular the issue of water stress, supply and scarcity (Brown & Matlock (2011:1).

According to Perreira (2017:34) The indicators and indices of water scarcity in addition to assessing and characterizing the water situation of the region or river basin are also used in order to mitigate the risk of drought in the long term, and the following indices and indicators are highlighted:

1.29.1 Falkenmark Water Stress Index (WSI)

This indicator, devised by Falkenmark in 1989, relates the water resources available in a given region or country per year to the number of inhabitants regardless of the temporal and spatial distribution of the water resource (Brown & Matlock, 2011:1).

Eq6: Water stress equation

$$fl(\text{m}^3 / hab) = \frac{\text{Recursos hidricos disponiveis (anuais)}}{\text{numero de habitantes}}$$

Where:

Fl - Falkenmark

m^3 - Cubic meters

hab. - Inhabitants

Quantified per capita water availability is then compared with overall indicative figures established by Falkenmark. In this region the defined categories are: no water stress, with water stress, with absolute scarcity.

Table 7: Falkenmark indicator for water scarcity

FI (m³/hab.)	Nível de *stress* hídrico
> 1700	Sem *stress* hídrico
1000 – 1700	Com *stress* hídrico
500 – 1000	Com escassez
< 500	Com escassez absoluta

Source: BROWN & MATLOCK (2011:1)

The ranges established by Falkenmark consider that water stress begins when per capita availability in a region is less than 1700 m^3/inhabitants and when the same value is less than 1000 m^3/ inhabitants. It is already considered that there is shortage (PERREIRA, 2017:18)

1.29.2 Social Water Stress Index

This indicator is the result of the change in the Falkenmark indicator where Ohlsson takes into account the ability to adapt socially to water stress through the use of technological, financial or other means that can meet the challenge of water availability. Thus, the social water stress index for Ohisson is based on the human development index, which implies that the capacity to adapt depends on the allocation of resources associated with the level or degree of education and political commitment (Nepomilueva, 2017:11).

1.29.3 Basic Human Water Requirements

Developed by Gleik (1996:88) is an indicator that allows to meet basic human water needs such as the consumption of water for survival, water for sanitation services and domestic use for food preparation. The minimum quantities of water required for basic human needs are as follows:

Table 8: Basic Water Requirements

Basic Water Needs	**Minimum quantities of water per capita/ per day**
Water for consumption	5 Liters
Water for environmental sanitation	20 Liters
Water for the Bath	15 Liters
Water for food preparation	10 Liters

Source: Author's elaboration according to BROWN & MATLOCK (2011:2)

The proposed indicator of basic human water needs totals a demand of 50 litres of water per day and per person regardless of climatic conditions, technology or culture.

1.29.4 Water Poverty Index (WPI)

Developed by Sullivan (2003) assesses water security for household use and other community needs such as agriculture combined with maintaining the integrity of the ecological balance (Rijsberman, 2004: 3). In the view of other researchers such as Feitelson & Chenoweth (2002:270) the water poverty indicator should incorporate the inability of a country or a region to afford the cost of drinking water for all the population at the same time which implies the need to find basic drinking water criteria and include the costs of treatment when necessary.

1.29.5 Water Footprint

Idealized by Hoekstra et al (2011:2) starts from the idea of considering the use of water along the productive chains, i.e. people use a lot of water for drinking and washing, even more for the production of food, paper, cotton clothes, etc. However the water footprint is understood as an indicator of the use of fresh water that analyses its use directly and indirectly in a temporal and spatial dimension.

The water footprint can be divided into three groups: the blue, green and grey water footprint.

The blue water footprint of a product refers to the consumption of blue water (surface and groundwater) along its production chain. The green water footprint refers to the consumption of green water (rainwater, as long as it does not drain) and the grey water footprint refers to pollution and is defined as the volume of fresh water needed to assimilate the load of pollutants from natural concentrations and existing water quality standards.

With everything the water footprint offers a more appropriate and broader perspective on how a consumer or producer relates to the use of fresh water systems and not as a measure of the severity of the local environmental impact of water consumption and pollution.

1.29.6 Virtual Water

The concept of virtual water was introduced by Tony Allan of Kings College London in the early nineties. It was first discussed internationally in 2002 in Delft and at the third World Water Forum in March 2003 in Japan and the concept continues to be discussed around the world. (www.limpoporak.com consulted on 15/06/2019)

According to Allan (2003:5), virtual water is understood as "*that which is used for the production of agricultural commodities and it can be extended to include the water needed to produce non-agricultural commodities, i.e., the water used to produce a good, product or service.*

1.29.7 Physical and Economic Water Scarcity Indicator (IWMI)

The issue of water scarcity is quite controversial, as other authors such as Barros & Amim (2007: 89) state that the quantity and quality of water resources under natural conditions depend on the climate and the physical and biological characteristics of the ecosystems that compose it, which indicates that any change in these components will alter the quality, quantity and time of water resistance in ecosystems. This position is debatable in the eyes of Risberman Apud Perreira (2017:13) as it is not easy to determine whether there is in fact water scarcity in the physical sense globally or whether the resource is available but should be used more appropriately.

The International Water Management Institute (IWMI) used a slightly larger scale water scarcity assessment across the world and considered in the analysis the share of renewable freshwater resources available for human needs (accounting for the existence of water infrastructure) in relation to the main water supply (Brown & Matlock (2011:8).

The analysis classified countries as physically water-scarce when more than 70% of river flows are withdrawn for agricultural, industrial and domestic purposes. Indicators of physical water scarcity include:

✓ Acute environmental degradation
✓ Reduction of groundwater
✓ Allocation of water that supports some sectors to the detriment of
  others.

IWMI assessed the global water resources study and mapped the regions indicative of little or no physical and economic water scarcity as shown in the figure below.

Figure 24: Illustrative map of physical and economic water scarcity

Source: BROWN & MATLOCK (2011:8)

With everything analysing the situation of physical and economic water scarcity in Mozambique according to IWMI it is concluded that the southern zone fits in the classification of physical water scarcity while the rest of our territory fits in the classification of economic scarcity.

1.29.8 Index Based on Water Withdrawal

According to Wu & Hui (2017: 7) This index was developed by Vorosmart et al in 2005, which also became known as local relativity index in water use and its formula is:

Eq7: Water scarcity equation based on water withdrawal

$$ISH = \frac{DIA}{Q}$$

Source: Wu & Hui (2017: 7)

Where:

D = Quantity of water withdrawn for domestic use

I = Quantity of water withdrawn for industrial use

A = Quantity of water withdrawn for agriculture

Q = Accumulated discharges along the river network

The ratios by the studies previously made by Raskin et al in 1999 where it is concluded that the water stress index starts when water withdrawal increases by 10% of Q which assumes that if ISH is greater than 0.2 there will be a limitation in economic development and it is 0.4 is considered high. See the table below

Table 9 Index category based on water withdrawal

Category	Index
Bass	<0.1
Moderated	0.1-0.2
Medio	0.2-0.4
Top	>0.4

Source: Raskin et al, (1997:28)

1.29.9 Water Supply Sustainability Risk Index

Developed by Rayal, where its values are calculated on the basis of the sum of the values of five criteria. His formula is:

Equation 8: Equation of sustainability risk in water supply

$$IRSOA = \sum_{i=1}^{5} Criteria_i$$

Each criterion has the value 1 if the value of a certain country is or exceeds the limit of this criterion and the opposite the marked value is 0 (Zero), thus the values of IRSOA can vary from 0 to 5 and that the high value represents greater risk.

Table 10: Representation of the category of Sustainability Risk in Water Supply

Category	Index
Precipitation	Precipitation available greater than 25%
Drought susceptibility	Water deficit in summer greater than 10% calculates on the basis of the differentiation between precipitation and water withdrawal in the summer period
Increase in water use	Total drinking water use increases its withdrawal by more than 20% between 2005 and 2050
Increased need for water conservation	Shortage in summer increases over index 1 from 2005 to 2050
Use of water from groundwater aquifers	The withdrawal of groundwater as a fraction of the total water withdrawal and more than 25%

Table 11: Classification of Water Supply Sustainability Risk.

Category	Index
Low Risk	<2
Moderated	2
Top	3
Extreme Risk	>4

Source: ROY et al 2012

1.29.10 Climate Index

Climate data are important in analyzing the socio-environmental vulnerability of water scarcity as temperature and precipitation are important elements in the abundance or scarcity of water in a given place (CC, 2014:15).

The most striking feature in Mozambique's climate is the very clear annual variation in the amount of rainfall that is associated with the ENSO phenomenon. The cold phase (La Nina) produces rainfall above the climatological average and the hot phase of this phenomenon (El Nino) produces rainfall below the climatological average and consequently drought, see the figure below, (INAM, 2002:15).

Figure 25: Time chart of La Nina and El Nino Phase in Mozambique (1978-1997)

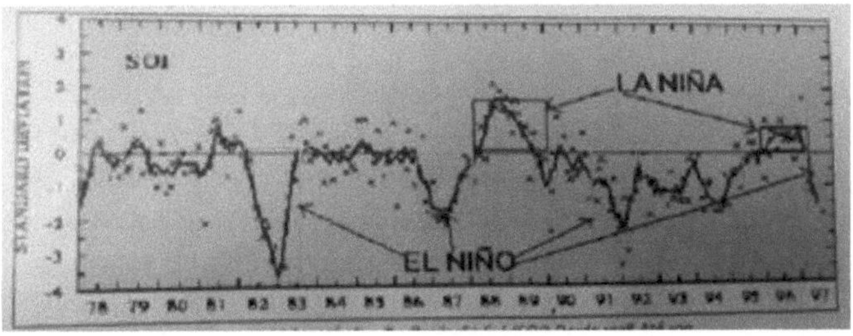

Source: INAM, 2002

1.29.11 Vulnerability Index on the basis of the Declivities

The slope is the degree of slope of a given terrain in relation to the horizon, the steeper the slope, the greater the direct runoff contributed to the decrease of infiltration, which means loss of precipitated water in high areas and accumulation of water in low areas. This variable helps in the analysis of the use of water resources in the study site.

1.29.12 Representative Table of Values of Flow Coefficients according to Soil Type, Declivity and Plant Cover.

According to Kobyama, (2008:48), due to the hydrological cycle there is spatial variability of water resources according to climate, vegetation and site characteristics (KOBYAMA, 2008:48).

Table 12: Flow Coefficient Values

Declividade (%)	Solo Arenoso	Solo Franco	Solo Argiloso
	Florestas		
0 - 5	0,10	0,30	0,40
5 - 10	0,25	0,35	0,50
10 - 30	0,30	0,50	0,60
	Pastagens		
0 - 5	0,10	0,30	0,40
5 - 10	0,15	0,35	0,55
10 - 30	0,20	0,40	0,60
	Terras cultivadas		
0 - 5	0,30	0,50	0,60
5 - 10	0,40	0,60	0,70
10 - 30	0,50	0,70	0,80

1.29.13 Vulnerability Index on the Basis of Soil Type Characteristics

The soil is the natural reservoir of water and the entrance of water occurs through the infiltration process which depends on the surface and structural conditions of the soil, mainly the quantity, size and continuity of the porous system (Reicheit et al, 2011: 34) The granulometry of the soil conditions its permeability, however, the finer the soil the lower its infiltration, this means that coarse-textured soils have less water retention capacity in the soil while heavy-textured soils have a higher water retention capacity. According to Goncalves et al (2016:16) the texture of soils is grouped as follows:

Table 3: Soil texture

Texture Grouping	Symbol
Coarse Texture	G
Average Texture	M
Heavy Texture	P
Medium Heavy Texture	MdP
Very Heavy Texture	MP

Author's elaboration based on the model of Goncalves et al (2016:16)

This grouping of soil texture is represented in the textural diagram according to the figure below.

Figure 26 Soil Textural Diagram

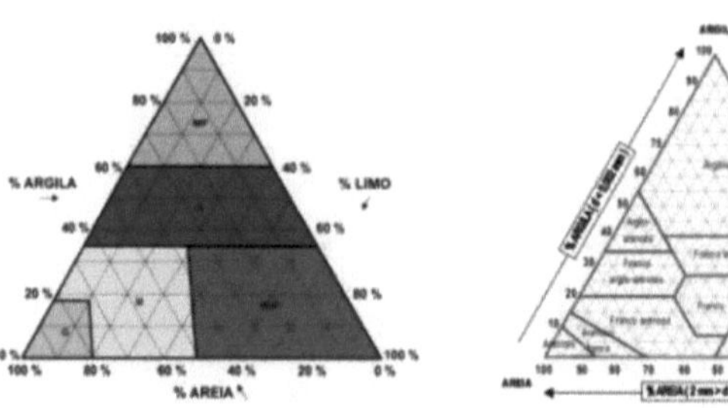

FAO Classification   US Classification

Using the triangular diagram for the classification of soil texture by the U.S. Department of Agriculture, the infiltration capacity of soils is represented as follows

1.29.14 Vulnerability Index at the Vegetation Base

Deforestation alters water and energy cycles, inducing an increase in air temperature and decrease in precipitation, which can reduce the amounts of water vapor exported to other regions. The fragmentation of forest areas decreases the number of habitats available for biological species (SALATI et al, 2006:109) on the other hand the density of vegetation cover is an important factor in reducing sediment removal, runoff, land surface erosion and soil loss (BOSETTI, 2010:32), (HIPOLITO et al, 340: 2011)

2nd Chapter

2.0 Physico-Geographical and Socio-Economic peculiarities of Funhalouro District

Figure 27: Map of Funhalouro district

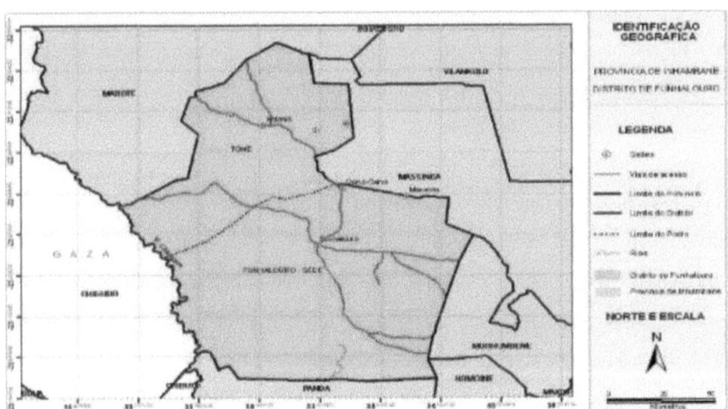

Source: PEDD, 2011

2.1 Location of Funhalouro District

According to PEDD (2011:3) the district of Funhalouro is cosmically located at 35°24' South Latitude and 33°35' East Longitude, having as limits, to the south district of Panda, to the east the district of Massinga, Morrumbene and Homoine, to the north the districts

69

of Mabote and some strips of Inhassoro and Vilanculos districts and to the west the province of Gaza.

The District of Funhalouro occupies an area of 13653 Km² (MAE, 2005:2)

2.2 Administrative Division of Funhalouro District

The Funhalouro District Station is composed of 4 locations as illustrated in the table below.

Table 4: Administrative Division of Funhalouro District Board

Administrative Post	Locations
Headquarters Administrative Post	Mucuine
	Mavume
	Manhiça
	Cupo
Administrative Post of Tome	Take
	Tsename

Source: GF, (2012:3)

2.3 Geology

The district of Funhalouro is characterized by a geological structure of fanerozoic rocks from the upper tertiary and lower quaternary periods with predominantly developed sedimentary formation of sandy and sandy-clay alluvium, conglomerate stoneware and limestone justifying the abundance of the stone the surface.

2.4 Relief

The district of Funhalouro is characterized by sandy plains and inland dunes with altitudes of less than 200 meters. This region is located on a levelled surface of accumulation origin and towards the East with a plain of extrusive origin stripped naked and some depressions of accumulation that condition the type of land occupation due to the scarcity of wetlands for the practice of agriculture (GF, 2011: 4)

2.5 Soil

In the district of Funhalouro predominates soils originating from mangrove sediments with sandy cover, from inland dunes in general derived from calcareous rocks, namely: clayey soils from rains and sandy soils in the dune phase. Alluvial soils also occur along the natural boundary of the Changana River with the province of Gaza as well as in the extreme south of the Mavume administrative post (ESF, 2010:27).

2.5.1 Sandy and Dune Roof Soils

They are depositions of the sedimentary basin occupying extensive areas of the district. The inland dunes are characterized by a gently undulating dune relief to the interior with good drainage, moderate to strong acids, low to moderate organic matter content, not saline and not sodium. These soils also have low water and nutrient retention capacity and low fertility, thus constituting limitations for agriculture but a good aptitude for forest development. (GF, 2011:4).

2.5.2 Mananga Sediments

These soils are distinguished in this region by the groupings M, MC, and M+MC, originating from clay translocation processes. They are sediments constituted by hard sodium deposits of Pleistocene in association with the sandy cover. These soils have the particularity that during the rainy season they present marshy depressions with a not very thick sandy cover. Often an extremely hard layer develops which makes it difficult for the roots of the plants to penetrate. The main limitation of this type of soil for the practice of agriculture is related to salinity, sodicity, drainage and the occurrence of periodic floods.

Figure 28: Soil map of Funhalouro district

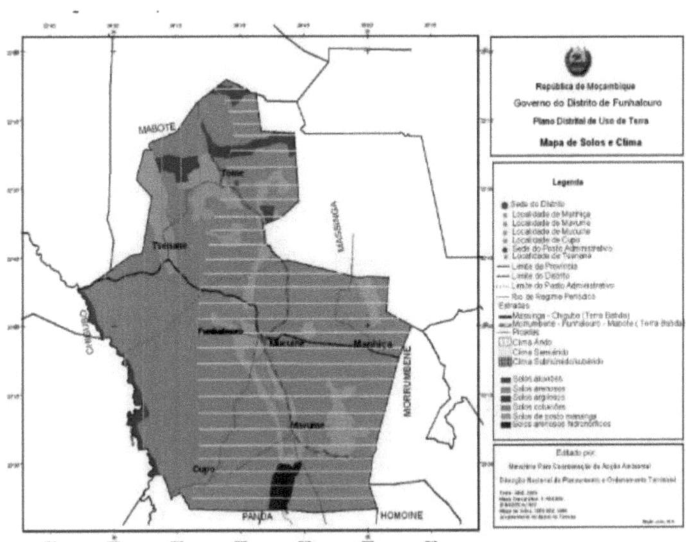

Source: CENACARTA, 2018

2.6 Hydrography

Climatic factors, the relief and the nature of the soils have a considerable influence on the flow, the structure and the pattern of the hydrographic network, contributing in this way to the recording of its maximums during the rainy season and the minimums during the dry season, with everything going through dry and permeable regions, the rivers lose a large part of their water either by infiltration or by evaporation recording the underground run-off.

This partly explains the water deficit that characterises some valleys and the surrounding regions, which is one of the greatest problems for the population and livestock, and that this situation is further aggravated by the occurrence of brackish waters at depth or on the ground of the river valleys in their terminal part (Dos Muchangos, 1999:46).

The district of Funhalouro in the southwest region above all in the plain has the presence of the tributary Changana River of the Limpopo River which serves as the natural border with the province of Gaza. The Changane River is a regular river and its flow is only formed during the rainy season and is not usable due to its high salinity.

Figure 29: Map of Hydrography, District of Funhalouro

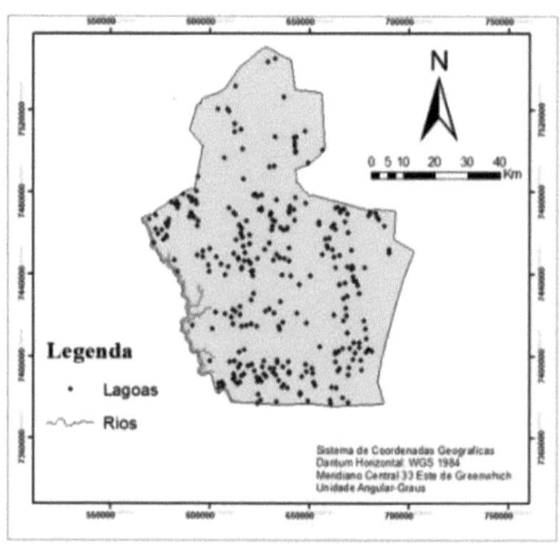

Source: Author, 2019

According to ESF (2010: 27) in the center of the plateau the groundwater of the Jofane limestone is saline and the biggest restrictions to the use of groundwater in this region are the poor quality of the water, the great variability of the productivity of the aquifer, depth and the great fluctuation of the level of the water table being more than 15 m deep. Although the district is not crossed by any river, there are some lagoons concentrated along the district boundary in the south and east areas (Mavume and Copo localities), in the villages of Mukhizame and Zivine, repectively, widely used as sources of supply for small irrigation systems, livestock watering and family consumption (GF, 2011:6).

2.7 Climate

The district of Funhalouro is dominated by a semi-arid tropical climate, i.e. hot and dry and cloudless in summer and cold in winter, with two distinct periods: dry and cool from April to November, usually the longest and hottest and rainiest from December to March, relatively short. This type of climate is characterized by average annual rainfall ranging from 500 to 800 mm, evapotranspiration exceeding 1500 mm, average annual temperatures above 24°C and associated with cyclonic, weak and moderate winds from the Northern Hemisphere.

Analyzing the temperature regime and the amount of precipitation and evapotranspiration, it appears that there is more evapotranspiration than rainfall, clearly justifying the shortage of water in the district **see figure n° 24** (GF, 2011:7).

Figure 30: Temperature and average rainfall graph for Funhalouro District (2010)

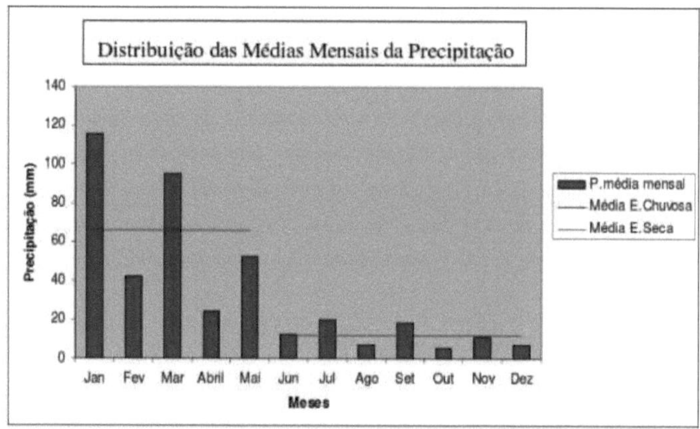

Source: INAM, 2012

2.8 Vegetation

The district of Funhalouro is mostly dominated by deciduous vegetation that characterizes the majority of the forests of this region with the greatest specific diversity well distributed in all localities, being the species of the acacia genus the most common (GF, 2011:5).

Figure 31: Vegetation Map of Funhalouro District

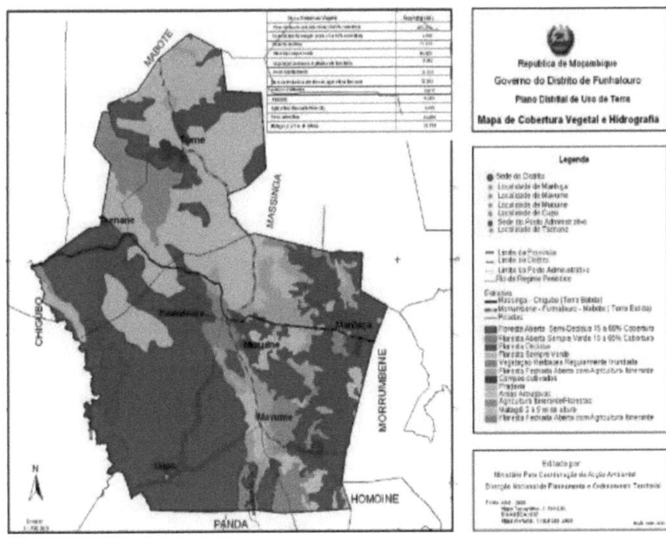

CENACARTA: 2011

2.9 Socio-Economic Aspect of Funhalouro District

2.9.1 Population

The District of Funhalouro according to INE (2007) has a population of about 37856 inhabitants mostly matswas distributed as follows:

Table 13: Population of Funhalouro District

	Total	Men	Women
FUNHALOURO District	37 856	16 788	21 068
FUNHALOURO Administrative Post	26 121	11 597	14 524
Locality MUCHUHUINE	15 508	6 783	8 725
Locality MANHICA	2 856	1 320	1 536
Location MAVUME	6 350	2 856	3 494
Location CUPO	1 407	638	769
Administrative Post TOME	11 735	5 191	6 544
Locality TOME - HEADQUARTERS	8 805	3 910	4 895
Locality TSENANE	2 930	1 281	1 649

Source: INE, 2007

2.10 Historical Overview of District Name

Around 1910 the district of Funhalouro was a small village called Mazive because it was inhabited by descendants of Mazive, a vassal of the Nguni army that at the time of great wars settled in the area to control the land. By this time a caravan that left Mocoduene walking Chinguendele towards Vilanculos passed through this area having a Portuguese delivering the caravan, fixing his gaze to the west side, glimpsed in the distance on the horizon what seemed to him a course of water.

Two men were sent to investigate the place and once there they found Chief Zequiane. As they were both starving, they asked him for food, to which he replied in the local language *"Oho Anzina tchumo hambo mina cima foolo a rumbo ga mina go Fonhololo hi mdlala"* which means: *"Oh! I have nothing. Even I since yesterday my belly is wrinkled with hunger"*. With the settlement of the Portuguese settlers in the region, they turned Fonhololo's pronunciation to Funhalouro.

In 1942, by Decree 31896 of 27 February, the district of Massinga was created and annexed to the territories of Mazive and Manhiça, at the time belonging to the district of Vilankulos. Although it was by this device, the Massinga constituency did not operate in the same year, partly due to the lack of budget allocation as its area of jurisdiction was not completely defined.

The fifteenth Ordinary Session of the Popular Assembly, held from 21 to 26 July 1986, approved the new administrative division of the country, creating Administrative Posts and new Districts.

It was in that year that Funhalouro ascended to the District Statute through Resolution number 6/86 of 25 July (GF,2011:15).

2.10.1 Religion

The population of the Administrative Post of Funhalouro professes several religions among which the United Methodist Church in Mozambique, Apostolic Faith Mission in Mozambique, Free Methodist in Mozambique, Apostolic New Bethel in Mozambique, Jerusalem of Mozambique, Zion House of God in Mozambique, Roman Catholic, Old Apostles among others (GF, 2011:16).

2.11 Socio-economic Infrastructures

To appreciate the causes of human distribution, the reasons for its thickening or rarefaction are not enough to confront it with natural conditions. The settlement where it exists has an exceptional artificial character due to an external artificial stimulus (trade, communication, use of raw materials that will be consumed elsewhere, military base, scientific observatory that limits its extent MASSINGUE, (2012:32).

The main housing areas in the district of Funhalouro are located near small water sources. Most of the houses are made of local material (stakes and grass) and mixed material (zinc sheets and stakes) as illustrated in the figure below.

Figure 32: Funhalouro District Housing Types Photo

Source: Author, 2019

The town of Funhalouro, is the area with the largest number of masonry houses, mostly belonging to public institutions, traders, loggers, miners and employees. The district also has a road network made up of tertiary and side roads, see figure no. 33, for a length of 878 km unpaved without a bridge since the district is not crossed by any river.

Figure 33: Photo of the road linking the village of Massinga with the district of Funhalouro

Source: Author, 2019

The water supply is made from boreholes and open wells in addition to lagoons making up a total of just over 100 water sources. The number of existing pumps is close to the population's needs, in addition to recording frequent breakdowns.

Figure 34: Photo of a well at Funhalouro, Tsename Administrative Post

Source: Author, 2019

The district's education system consists of 36 educational establishments, namely 1 ESG of 1° Cycle, 11 EPCs, 24 EP1, 27 attached rooms and 102 literacy and adult education centres.

Figure 35: Photo of a school, Mucuine Administrative Post

Source: Author, 2019

As far as the health network is concerned, the district consists of 4 health centres distributed as follows: Funhalouro Administrative Post has 5 conventional Health Centres of which two are in a good state of conservation and one (1) in a state of degradation in the village headquarters of Funhalouro and Posto Administrativo de Tome have 2 conventional health centres in good condition.

Figure 36: Funhalouro Health Center photo

Source: Author, 2019

The District of Funhalouro also has mobile phone services from Mozambique, MCEL, Vodacom Mozambique, VODACOM and Movitel, therefore it also has services, internet, radio communication, BCI Bank and fax, however the access to these services are very poor since they are only available at the headquarters of the Administrative posts with a radius ranging from 10 to 15 km.

Figure 37: Photo of the First Commercial Bank in Funhalouro's head office village

Source: Author, 2019

2.12 Socio-Economic Activities

Agriculture is the dominant activity and involves almost all households and this activity is fundamentally non-irrigated and the crops are done by hand. In all localities of this district maize, peanuts, Nhemba beans, sorghum, cassava, yogh beans, cotton and cashew

79

are cultivated, but despite their importance they face many difficulties for their marketing because they are areas of difficult access.

In this sector of activity, Agriculture, the district is challenged by the irregularity of rainfall, the great vulnerability to natural disasters that condition the potential of agricultural production of dryland

In the livestock area, despite the deficiency in the veterinary assistance of the animals, the district presents great economic potential given the existence of good grazing areas.

Figure 38: Photo of the Agricultural Space, Mavume Administrative Post.

Source: Author, 2019

The flora of the district on the other hand is a great economic opportunity in species such as chanfuta, Umbila, Mecrusse, Missassa branca, among both the forest resources of the district are exploited for logging, stakes for construction and sale to the public, firewood and charcoal.

III Chapter: Presentation and Analysis of Water Scarcity Response Data in the District of Funhalouro

This study dealing with semi quantitative approaches that integrate vulnerability factors and vulnerable elements considers that the perspective of vulnerability analysis can result in a social and/or spatial hierarchy of the exposed elements and that this requires measurements and representations based on both environmental and social aspects because they complement each other. In this perspective, after studies on the main methods of water scarcity assessment, presented in the theoretical basis, the IWMI method was adopted due to its general approach in classifying water scarcity on the physical (environmental aspects) and economic (use of technologies) aspects in relation to other methods that focus only on physical water scarcity or economic water scarcity as illustrated in the table below.

Table 5: Illustration of indices based on physical and economic water scarcity

Methods based on physical water scarcity	Methods based on economic water scarcity
Falkenmark Water Stress Index (WSI)	Social Water Stress Index
Basic Human Water Requirements	Water Poverty Index (WPI)
Index Based on Water Withdrawal	Water Footprint
Water Supply Sustainability Risk Index	Virtual Water
Climate Index	Physical and Economic Water Scarcity Indicator
Vulnerability Index on the basis of the Declivities	
Vulnerability Index on the Basis of Soil Type Characteristics	
12 Vegetation Base Vulnerability Index	

Source: Author, 2019

In order to facilitate the understanding of the socio-environmental vulnerability of water scarcity the data are presented in a separate way, i.e. environmental vulnerability and then social vulnerability and finally the socio-environmental vulnerability that deserved an analysis in its global context.

3.1 Environmental Vulnerability to Water Scarcity

3.1.1 Lithological Factor

According to Mandala apud Crepani (2016: 155) the lithological units are understood as an important element in the analysis and definition of the morphodynamic category of the unity of the natural landscape and covers the resistance of the rocks that compose it understood as links between minerals or constituent particles.

According to ESF (2011:44) the constraint of drinking water availability in the district of Funhalouro is due to the strata that make up the subsoil of the district that are conveyed the maritime transgressions that occurred during the creation of the current formations left salt capes which leads to the occurrence of high content of salts in the groundwater, especially in the Mazamba tertiary plain and Urronga tertiary plain member of the Jofane formation.

Figure 39: Map of Water Vulnerability according to lithological factor

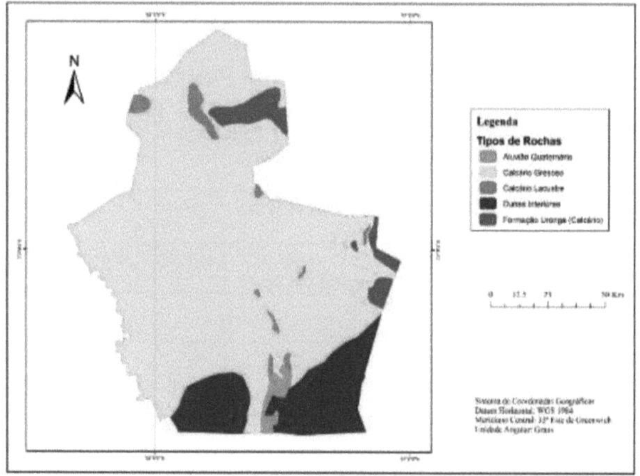

Author's elaboration based on ESF data, 2011

There are several microchemical and microbiological studies made in this district to determine the quality of water, this work highlights the research of Enginyeria Sense Fronteres et al (2010), Herculano's master's degree research (2012), INE's research (2012) and DNA research represented in the hydrogeological map of Mozambique on the scale of 1:8000,000, 1992. These studies based on the chemical, physical organoleptic parameters of water for human consumption established by MISAU See Annex 1 and 2, it was concluded that sodium chloride varies from mild to high level in this district.

Figure 40: Sodium chloride sample in water collected from some wells in the district.

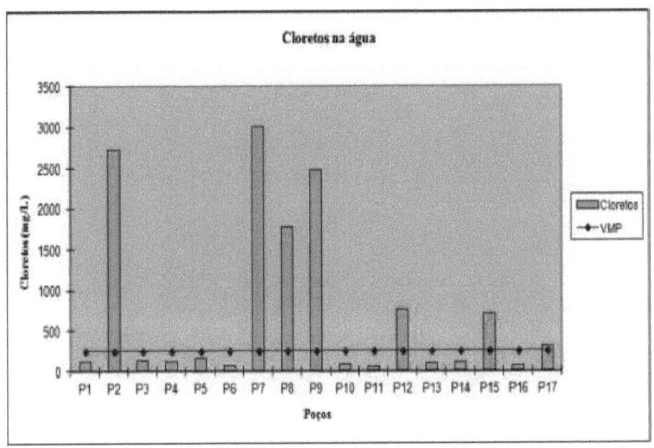

Source: HERCULAN, (2012:57)

Statistics Portugal has reached the following conclusion in its surveys:

Relatório do Distrito Funhalouro

16	Zonas de risco Hidrogeológico

O SDPI indicou, baseado na experiencia adquirida, quais são os problemas hidrogeológicos na zona. Os valores indicados devem ser considerados como indicativos.

Classificações segundo SDPI				
Posto Administrativo	Profundidades normais dos furos	Taxa de sucesso para abrir furos com água	Qualidade de água (salinidade)	Possibilidade para abrir poços
Posto Sede	50-75m	60-80%	Muitos fontes salobras	95-100%
Posto Adm.Tome	50-75m	60-80%	Muitos fontes salobras	80-95%

Source: INE, 2012

With regard to other analysis results it was found that the values are within the normal range, e.g. nitrate (> 10 mg/l), ammonia (> 1.5 mg/l) and Ph (minimum 5.98 and maximum 7.93) turbidity (varies from 0.1 to 3 NTU).

According to E. F. 1, water scarcity in the district of Funhalouro is not only justified on the one hand by the salt cliffs left in the rocks during the period of marine regression, but on the other hand

The soil being made up of sandstone and conglomerates makes it difficult to drill wells and boreholes, and the district government does not have the

financial capacity to cover the costs of drilling operations at district level, and
this situation is further aggravated by the lack of rain.

Figure 41: Stoneware and conglomerate blooms on the surface, Funhalouro headquarters.

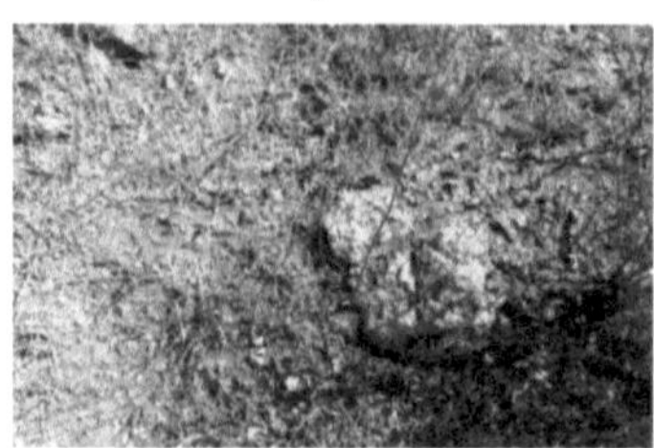

Source: Author, 2019

The question of the availability of saline water in the district of Funhalouro for reasons of salt layers in groundwater is debatable because according to E.F.2

The perforation of the water holes should go beyond the limestone layer and
the perforation depth should vary from 60 meters to 100 meters, at these depths
you will find water with less salt content than Funhalouro water.

The procedures used to find water with lower salt content, according to E. F. 2 it is necessary to do the drilling and in the salt water towel it is necessary to do the insulation of pipes and to put separators of the salt water with the fresh water as you continue with the drilling, but this costs a lot of money.

The experience of this operator proves the veracity of the study made by DNA (1992) sees annex n° 3 that refers to the existence of regions with brackish water over fresh water and regions of fresh water over brackish water, that is, there is a semi-confined or suspended aquifer that is defined according to Paiva et al (2001) as the water that infiltrates the ground and encounters an obstacle, an impermeable surface that impedes its descent to the water table. For this reason, the differentiation in salt contents between the two aquifers may occur, as illustrated in the following image of two fountains that are approximately 100 meters apart in MUCUINE

Figure 42: Highly salted water   Lightly salted water

3.1.2 Factor of Declivity

Figure 43: Slope map of Funhalouro District

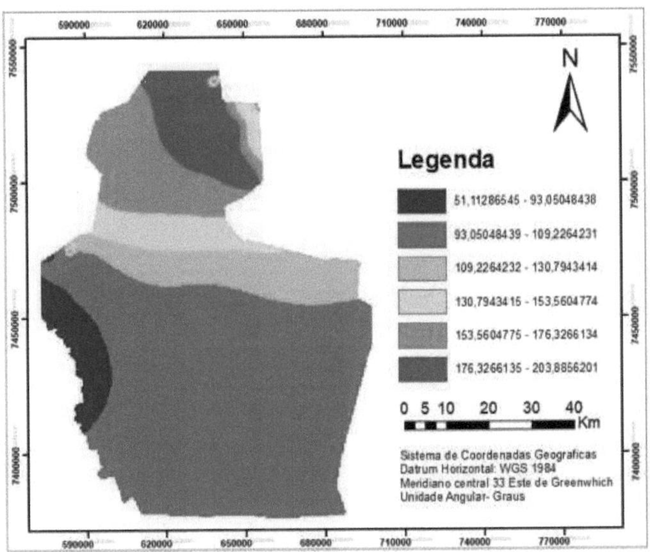

Source: Author, 2019

The orography of the Funhalouro district is well differentiated between the southern part, lower and the northern part, higher. The southern part is like a plain about 80 m wide divided into two parts by a central depression that shows the bed of an ancient river, where its main geology is alluvial (thick material not consolidated) and the northern part is

85

higher (150 meters on average) although there are no mountains or considerable slopes, see the figure below.

With regard to vulnerability to water scarcity according to E.F 1 reiterates the following:

The northern zone and one percent of the district is more vulnerable to water scarcity than the southern zone due to the configuration of the relief and the depth of the aquifers, but in the southern zone what aggravates the lack of water most is the scarcity of rain because if the rain occurred regularly the lakes and lagoons would have water to supply the population.

Figure 44: Dry lake for lack of rain

Source: Author, 2019

But from the point of view of public health safety, according to E. F. 3

The occurrence of rain in the district, brings several health problems because in the rainy season there are more cases of acute diarrhea, viruses, scabies and there is a record of deaths although in smaller numbers due to the consumption of water from lakes and ponds without proper treatment for human consumption.

This position leads to the conclusion that, although low-lying areas are less vulnerable to water scarcity in the rainy season, they also constitute areas at great risk to public health in violation of Article 8(b), see Annex 4

However, although there are no mountains or considerable slopes in the district of Funhalouro for the construction of a dam, article 63 of the water law recommends the creation of rain protection areas for the reservation and maintenance of aquifers, which in this case could include low areas especially for lakes and lagoons in the district.

3.1.3 Soil Salinisation Factor

According to De Lima et al (2010:3) and FAO: (2011:120) Soil salinization is one of the growing phenomena worldwide, mainly in arid and semi-arid regions, severely limiting agricultural production and is defined according to Goncalves et al (2009:3) as a process that leads to increased concentration of soil solution in soluble salts (Na^+, Ca^+, Mg^+, K^+) to levels detrimental to plants.

The genesis of soil salinisation has a high relationship with both the predominant geological formation in the landscape and the drainage

According to Pedrotti (2015:1310) The excess of soluble salts in the soil solution is the result of a combination of climatic factors (low rainfall and high evapotranspiration rate), soil conditions (low leaching capacity of salts and presence of impermeable layers) and soil management (irrigation with saline water)

Figure 45: Vegetables irrigated with slightly salty well water, accumulation of salts in the soil.

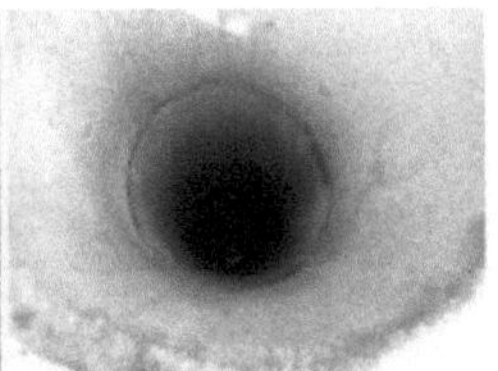

Source: Author, 2019

However, the combination of climatic and geological factors gives the district a variation in the salinisation content of soils from mild to moderate as illustrated in the tables below according to administrative posts, but there is a high risk of strong salinisation of soils due to high sodisation (GONCALVES, 2009:3)

Administrative Post of Copo and Mavume

These administrative posts are characterised by very deep yellowish brown sandy soils on a terrain with inland dunes interrupted by managerial soils. The drainage of this soil varies from good to moderate excessive and has a low vulnerability to water scarcity due to high to medium water infiltration capacity and salt leaching, see figure n° 26 triangular diagram for soil classification and infiltration capacity.

Table 14: Soil characteristics of the Administrative Post of Copo and Mavume

Soil group	Layers	Salinisation % Salinisation	Sodicity % Sodicity	Acidity % Acidity
dA+M	Surface	Salted on 0.5-7.5	Non-sodium 0-2	Strong 4-6
	Underground	Moder. salty 3-15	Light sodium 1-12	Strong 4-5

Source: Author's elaboration based on data from the Mozambique soil charter

Mucuine and Manhiça Administrative Post

Characterized by dark greyish brown mangrove clay soils in a circular depression terrain, it presents an imperfect drainage to bad reducing in this way the infiltration of water into the water table, thus constituting a vulnerable area to water scarcity, see figure no. 26, triangular diagram for soil classification and infiltration capacity.

According to the table below, the salinization and sodicity of this administrative post are the main limitations for the practice of agriculture and although it has a moderate salty subsoil, it is at risk of being strongly salty due to strong sodicity ranging from 10 to 40%.

Table 15: Soil characteristics of the Mucuine and Manhiça Administrative Post

Group/soil	Layers	Salinisation % Salinisation	Sodicity % Sodicity	Acidity % Acidity
M+MC	Surface	Salted on 0.5-7.5	Light sodium 1-8	Mod. 5-7
	Underground	Moder. salty 3-15	Strong sodium 10-40	Fort 5-8

Source: Author's elaboration based on data from the Mozambique soil charter

Administrative Post of Tome and Tsename

The soils of these administrative posts are red clay derived from limestone, presents an imperfect to moderate drainage making water infiltration difficult. The similarity of the Manhiça administrative post, the perception of salinity and impermeability are limitations for the development of agriculture.

Table 16: Soil characteristics of the Administrative Post of Tome and Tsename

Group/soils	Layers	Salinisation % Salinisation	Sodicity % Sodicity	Acidity % Acidity
MC+M Wv	Surface	Salted on 0.5-7.5	Light sodium 1-8	Mod 5-7
	Underground	Moder. salty 3-15	Strong sodium 10-40	Fort 5-8

Source: Author's elaboration based on data from the Mozambique soil charter

Funhalouro Administrative Post

It is characterized by clayey mango soils and presents an imperfect drainage to bad reducing the infiltration of water into the water table thus constituting a vulnerable area for water scarcity

Table 17: Soil characteristics of the Funhalouro Administrative Post

Group/soil	Layers	Salinisation % Salinisation	Sodicity % Sodicity	Acidity % Acidity
MC	Surface	Unsalted 0,5-2	Light sodium 1-8	Mod. 5-7
	Undergroun d	Salty 0.5-9	Strong sodium 10-40	Fort 5-8

Source: Author's elaboration on the basis of the Mozambique Soil Charter.

The salinization of soils in the district of Funhalouro and the combination of water scarcity and rain damages agricultural activity, a sector with the highest demand for water in the district according to the population in question, puts the food security of the population at risk.

Figure 46: Graph of Water Demand in Funhalouro District

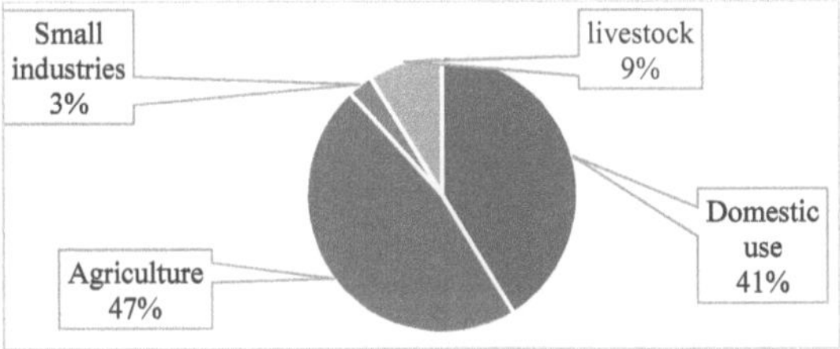

Source: Author, 2019

3.1.4 Climate Factor

The geological facts of the district of Funhalouro explain in part the reasons of the salinity of the soils and consequently the existence of the saline water, but in the occurrence of this phenomenon there is a need for the combination of the thermo pluviometric indexes of the district whose constitute catalysts of the process of salinization insofar as the average annual temperatures are **above 26 °C, cause high** evapotranspiration and consequently low precipitation which has the impact of reducing the recharge capacity of aquifers and the leaching of salts contained in the soil, thus putting at risk the availability of fresh water in the study site. See the graph below.

Figure 47: Graph of the Thermopluviometric Representation of Funhalouro District (2000-2017)

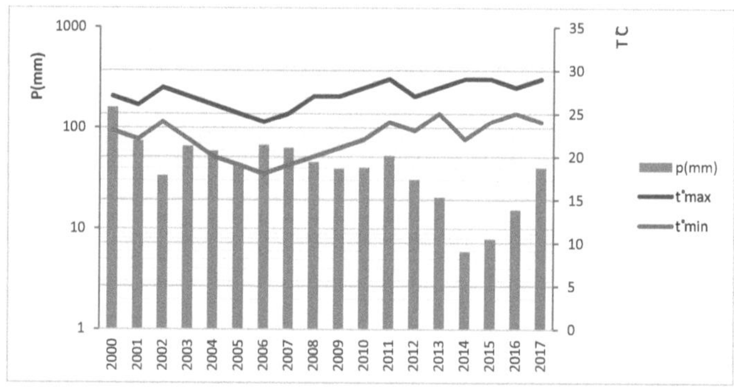

Source: Author, 2019 based on INAM data

This figure shows that in 2011-2014 there is a fall in rainfall due to climate change, with a strong focus on the phenomenon of el Nino, which represents an increase in temperature and a decrease in rainfall, providing long periods of acute drought, which on the other hand conditions the increase in the concentration of salts in the soil, damaging agricultural production.

3.1.5 Vegetation Factor

Although the relations between water production and forests are not a consensus from a scientific point of view, it is undeniable that the degradation and scarcity of the one impairs the existence of the other, according to Whately et al, (2008):27) river basins with vegetation cover have a greater contribution to the production of good quality water than others altered by different human activities and with different levels and types of contamination and on the other hand according to Mills et al (2002) there is a perception that forests act like sponges, absorbing water and gradually releasing it in times of drought, but in places where deforestation is associated with high soil compaction, surface runoff can increase in greater proportion and there is a decrease in evapotranspiration as a result of deforestation, generating a decrease in the level of the water table.

Figure 48: Photo of Funhalouro Vegetation

Source: Author, 2019

Other researchers such as Nucci, (2008:23) cite the importance of vegetation as surface stabilization through soil fixation by the roots, protection of springs and springs.

Besides the ecosystem service of water purification, the forest floor is formed by a layer of leaves, branches and other plant remains that provides it with great roughness, preventing the superficial run-off of water to the lower parts of the ground, favoring the infiltration of a portion of water that contributes to the gradual increase of the water table through subsurface run-off.

However, the destruction of the vegetation cover in Funhalouro through uncontrolled burning and the extraction of firewood and charcoal for commercialization damages in some way the ecosystem services of the thermal balance and being a semi-arid zone it becomes more vulnerable to climate changes with a greater focus on drought.

Figure 49: Photo of uncontrolled burning at Funhalouro

Source: Author, 2019

Figure 50: Photo of the extraction of firewood and charcoal for commercialization

Source: Author, 2019

It is necessary to stress that based on vegetation one of the old water storage strategies in this district was done in the surrounding areas where the tree was dug to collect rainwater.

Figure 51: Embondeiro photo, formerly used for rainwater harvesting and storage

Source: Author, 2019

Although the role of the embondeiro in water conservation is recognized, the population knows very little about the role of vegetation for the maintenance of ecosystems as well as its connection with precipitation and purification of water as illustrated by the figure in chart no. 52.

Figure 52: Graph of the population's perception of water scarcity in the district of Funhalouro

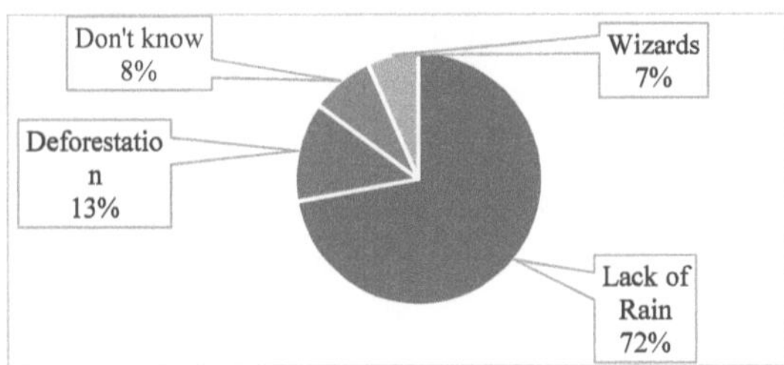

Source: Author, 2019

As a strategy to mitigate *the* impacts of deforestation the district government according to E. F. 1 "*has a reforestation plan although it does not use native plant species of the district, so it intends to plant approximately 1000 cashew trees.* "

Figure 53: Picture of the Nursery with cashew tree seedlings

Source: Author, 2019

3.2 Social Factor

Lack of access or poor access to water and sanitation are at the root of serious social problems as water scarcity and poverty are closely linked, thus violating the fundamental basic needs of the population in terms of health, food security and that each year water-borne diseases affect more than five million people in the world of which about 1.8 billion children die as a direct result of diarrhoea and other diseases caused by polluted water and poor sanitation (LEITAO, 2009:118).

The deficit access to water in Funhalouro on the other hand according to E. F4 "*has to do with the constant breakdowns of some hand pumps for water collection*".

93

Figure 54:Photo of inoperable water sources

Source: Author, 2019

This situation of constant failure of hand pumps of water according to E.F 1 must be on the one hand of the forms of its management because,

After the construction of the hand pumps they are handed over to the community management for maintenance and repair in case of breakdown, but the community does not have the financial capacity for maintenance and repair of the hand pumps for water collection, this constraint causes the population to move large distances in search of operational pumps for water.

According to INE data from 2007, the population of Funhalouro in terms of professional occupation is characterised as follows:

Table 18: Occupation of the population of Funhalouro

Occupation	No. of dwellings	Scale of vulnerability	Vulnerability Level
Workers	10985	0,2	Bass
Unemployed	26808	0,7	Top
Total	37793		

Source: Author based on data from INE, 2007

The price of water in the district of Funhalouro according to E.F1 varies according to the table below

Liters of water	Price	Administrative Post
20 Liters	2, 00 Meticais	Funhalouro headquarters, Cupo and Mavume.
20 Liters	20,00 Meticais	Muchuine, Manhiça and Tsenami

The private operators according to E.F.2 charge *about 3000,00 Meticais in the water contract and still charge 40,00 Meticais for each m³, and the population needs at least 80 liters daily for domestic use, in a family aggregation of 7 people.*

For these economic reasons, 85% of the population surveyed in a sample of 180 people at a ratio of 30 people per Administrative Post in the district, uses water holes to obtain this precious liquid as illustrated in the graph below and see Appendix No. 4

Figure 55: Funhalouro District Water Access Chart

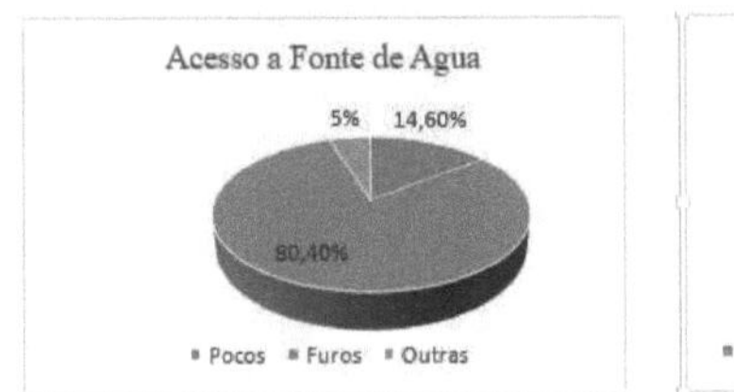

Source: Author, 2019

However, it is confirmed that most of Funhalouro's population is vulnerable to water scarcity because they do not have the financial resources to maintain and repair hand pumps for water collection as well as they are unable to access water supplied by private water supply operators, thus using wells and few water holes that are still operational.

Figure 56: Photo of Women seeking water in one of the fountains of Mucuine P.A.

Source: Author, 2019

And because other villages are relatively far away, about 5 km from the operational hand pumps, they use donkeys so that they can carry considerable amounts of water

Figure 57: Photo Equus asinus (Donkey) used for water transportation

Source: Author, 2009

Faced with this reality, according to E.F.1, a strategy has been adopted consisting of a rainwater collection system in case of rainfall, as this system is low cost, and is contributing to reduce demand for this precious liquid, although on the other hand, the risk of contamination of rainwater by corrosion of zinc plates and evaporation of water in the cistern is anticipated.

Figure 58: Photo of Rain Harvesting and Conservation

Source: Author, 2019

This strategy on the other hand faces challenges as most dwellings are haystacks made of piles and grass being inefficient for collecting more rainwater as illustrated in Table 19.

Table 19: Distribution of Housing Types by Population Number

Types of Housing	No. of dwellings	No. of persons	Scale	Vulnerability
Palhotas	6345	28506	0,8	Top
Conventional houses	14	109	0,001	Download
Other houses	1683	9178	0,2	Download
Total	8042	37793		

Source: Author's elaboration in INE database, 2007

This table shows that most people live in haystacks and have little chance of capturing rainwater for storage in cisterns, so when rainfall occurs they resort to lakes and small openings made along the road for rainwater drainage.

3.3 Institutional Factor

As a strategy to provide water during the drought period according to E. F. 1 the district government of Funhalouro in partnership with ARA-SUL built water dams having as priority the population, livestock watering and irrigation of the fields.

Figure 59: Photo of Water Dam, Thirsty Funhalur.

Source: Author, 2019

The water is pulled from the reservoir by a motor pump to a small water treatment centre where it is stored in a 5000 litre taque and then used for various purposes.

Figure 60: Photo of the Water Treatment Center Water conservation Taque

Source: Author, 2019

With everything there are major constraints that are encountered in these infrastructures and the problems are on the one hand of an economic nature and on the other hand of a technical nature, according to the explanation of E. F. 1 to the water treatment station faces a breakdown and has no specialized technician to repair and on the other hand the district has no funds to cost the repair of this equipment. For these reasons the dams are fenced for public use as their water does not benefit from any treatment. With everything there is a record of one death, a child suffering from thirst decided to break the fence that protects the dams in search of water and did this slipped and fell into the dam having died by drowning in the Administrative Post of Mucuine.

According to E.F.1 the number of SDPI employees from 2011 to the date of this survey had not yet known any change according to E.F.1, see table No. 17 and in addition to these employees the district has 3 mechanics for the maintenance of pumps.

Table 20: Number of Funhalouro SDPI Employees

Número actual de funcionários do SDPI (dados 2011)	Total	Departamento				
		Planeamento	Obras Públicas	Gestão Ambiental	Recursos Humanos	Outro
Técnico superior	1	1	0	0	0	
Técnico médio	1	0	0	0	0	
Técnico básico	0					
Ensino elementar	0	Sem Dados				
Número total	1	1	0	0	0	

Source: INE, 2011

However, the lack of human capital in quantity and quality as well as the lack of financial resources contributes to the district's hydrological scarcity as the hydrological infrastructure did not benefit from regula r maintenance.

Due to the institutional inability of the district government to provide water, the health centre in Funhalouro for example uses private operators to obtain water in order to develop its activities.

Figure 61: Ambulance seeking water from a private operator to Funhalouro Health Center

Source: Author, 2019

For the possibility of contamination of aquifers in the district see Appendix No 01

3.4 Hazards of Water Vulnerability in Funhalouro District

Exposure Elements	Danger
Population	✓ Acute diarrhea, ✓ Viruses, ✓ Sarna ✓ There's a death toll, ✓ Food insecurity
Cattle	Loss of heads of cattle
Excessive exploitation of forest resources	✓ Loss of plant cover, ✓ Increasing the level of risk to aquifer contamination, see Appendix No 3 ✓ Climate change

Source: Author, 2019

Conclusion

Water is an indispensable resource in the climate regulation of regions as well as for the survival of man and other living beings, but all over the world this resource is being degraded either by natural or human factors.

For the case of the district of Funhalouro the natural factors are pointed out as causes of water shortage, among them the lithology due to the existence of salts left during the period of marine regression which contributes to the salubrity of the water, the issue of uncontrolled burning having as impact the destruction of the vegetation cover and the compaction of the soil making it difficult to infiltrate the rainwater to feed the underground aquifers. On the other hand, there are climatic changes that influence the decrease of precipitation and increase of temperature, which accelerates the salinization processes of soils and the lower capacity to leach salts into the soil.

In terms of human factors, the lack of financial and economic resources for the expansion and maintenance of the water distribution network in this district contributes to water scarcity, forcing the population to travel about 20 km in search of water and it is constantly on the alert of food insecurity due to low agricultural production as a result of cyclical and prolonged droughts.

As a strategy for mitigating the effects of water scarcity, the population implements techniques for capturing and storing rainwater in small and medium cisterns and the institutional sector implements the inclusion of the private sector for water supply to communities.

However, although there is a water scarcity mitigation strategy in Funhalouro district, a large part of the population is marginalized in the implementation of these strategies due to lack of financial resources for the construction of improved houses and their cisterns as well as the purchase of water from private operators.

With this reality the capacity to respond to the vulnerability of water scarcity in the district of Funhalouro is still relatively low from the point of view of the use of the IWMI method competing for this analysis the demographic increase that has been putting pressure on water resources, environmental degradation through deforestation, uncontrolled burning, climate change, salinization of soils and institutional incapacity in the construction, maintenance and repair of water infrastructure, despite the implementation of strategies for capturing and storing rainwater.

However, in order to face the problem of water scarcity in the district of Funhalouro, it is recommended to encourage the population in practices of environmental conservation in relation to the conservation of forest species so that they contribute with their ecosystem value to increase the recharges and water quality of aquifers, thus combating soil salinization and reducing the impacts of climate change, it is also necessary to encourage the use of simple methods of water treatment such as the disinfectant called certainty, the use of moringa among other methods such as boiling water for domestic use.

Recommendations

Short Term Mitigation Measures

> ➤ To raise community awareness for the preservation and sustainable use of forest resources through the dissemination and implementation of actions against uncontrolled burning which alters the environmental quality and reproduction of aquifers
> ➤ Encourage the use of simple water treatment methods such as boiling water for domestic use, solar water desalination and the use of the disinfectant called certainty
> ➤ Deconcentrate the manual water pumps in the small villages of the Administrative Posts of this district in order to reduce the pressure and constant failure of the pumps which will contribute to the shortening of the distance in the search for water
> ➤ Guarantee the supply of spare parts

Long Term Mitigation Measures

Despite these challenges the district has several opportunities once being rich in limestone and stone one could set up a cement production plant with all the precautions of the environmental impact and the extraction of the stone for its commercialization in order to raise the GDP of this district as well as to alleviate the dependence of the agricultural activity which could help in the one:

> ✓ Creation of a fund for the management of semi-arid areas, focused on the construction of more hydraulic infrastructures (manual water pumps, dams, cisterns) for human consumption and livestock watering
> ✓ Allocation of human and financial resources for the maintenance of hydraulic infrastructures
> ✓ Allocation of water desalinators

There is a need for community-consciousness for the preservation and sustainable use of the priceless forest resources of the dissemination and implementation of actions against uncontrolled burning which alters the environmental quality and reproduction of aquifers

Bibliography

AIM. Shared Rivers reduce water flow to Mozambique, 2009, 3p

ALLAN, T. Virtual Water. The water, food and trade nexus, useful concept or misleading metaphor in IWRA- Water international V.28, London, 2003,11p

AMORIM, Miriam Cleide C. Environmental Pollution and Types of Pollution. UNVASF, Pernambuco, Brazil, 2011, 21p

WORLD BANK. National Assistance Strategy for Water Resources in Mozambique: Making Water Work for Sustainable Growth and Poverty Reduction, Maputo, 2007, 43p

BARROS, Fernando Gene Nunes & AMIM, Mário M. Agua: An economic asset of value to Brazil and the world, G&DR Editora, V.4, Brazil, 2007, 108p

BATOUXES, Maria & VIEGAS, Julieta. Dicionário de Geografia, Silabo, Lisboa, 1998

BIGAS, Harriet et al. The Global Water Crisis: Addressing an urgente security issue, Axworthy Editora, Canada, 2012

Bulletin of the Republic of Mozambique. Water Law no. 18/91 of August 3rd

Bulletin of the Republic of Mozambique. Regulation on water quality for human consumption: Ministerial Diploma n 180/2004 of September 15

Bulletin of the Republic of Mozambique. **Resolution n°53/2004 of December 1st**

BOOTH, Andrea et al. State of the Environment In Southern Africa, Creda Gauteng, Harare 1994

BOSETTI, Elvino Pinto. Geomorphology, UEPG, Brazil, 2010, 94p

BROWN, Amber & MATLOCK, Marty D. A review of Water Scarcity Indices and Methodologies, The sustainability consortium, USA, 2011, 17p

BURTON et al. Environment Hazards: Assessing risk and reducing disaster, 4ª Edição, Bell and Bain Ltd, Glasgow, 2008

CAMPOS, José Nilson Bessera. A Sustainable Development Strategy for the Northeast: Vulnerability of the semi-arid to droughts from the perspective of water resources, Brazil, 1994, 52p

CARCEDO, Francisco J. A. & JORGE, Olcina C. Riesgos Naturales, Ariel Ciencia Editora, Spain, 2002

CARDONA, Omar Dario A. La Necessidad de repensar de maneira holística los conceptos de Vulnerabilidad y riesgos: Una critica y una revision necesaria para la gestión, Holanda, 2001

CASTRO, M. Cleber et al. Environmental Risks and Geography: Concepts, Approaches and Scales, Vol 28, UFRJ Editora, Rio de Janeiro, Brazil, 2005

HORSE, José. The Drought in Northeast Brazil: A vision of study and research elaborated in a century of knowledge production, Fortaleza, 2008, 126p

CC- WARE, Mitigating Vulnerability of Water Resource Under Climate Change, In WP3 Report, UE, 2014, 70p

Centre for Weather Forecasting and Climate Studies (CEPTEC), 2010, available at http://enos.ceptec.ip.br/. Consulted on 15/09/2018

CERRI, LES & AMARAL, C.P. Geological Risks: Engineering Geology, São Paulo, Brazil, 1998

CHARTRES, Colin & WILLIAMS, John. Can Australia overcome its Water Scarcity Problems, In: Journal of Development in Sustainable agriculture, Sydney, Australia, 2006, 24p

CORREIA, Arlindo. Mozambique and the Water Sector: Development within the institutional framework of the water sector in Mozambique, P3IP, Figueira da Foz, Portugal, 2017, 19 p

CORREIA, Francisco Nunes & HENRIQUES, António Nuno Fernandes Goncalves. Convecções Sobre Agua, IST editora, Portugal, 2010, 25p

COUANA, Enoch Augustus. Influence of Enso on Precipitation in the Cities of Maputo, Beira and Lichinga: Graduation Work, UEM, Maputo, 2015, 46p

CUNHA, Ana et al. Availability of Freshwater: Will there be enough fresh water for human needs on this planet? U. Porto Editora, Portugal, 2009, 31p.

CUNHA, Lucius. Vulnerability: the less visible face of the study of natural risks, University of Coimbra editora, Coimbra, Portugal, 2006, 167p

CUTTER, Susan L. Vulnerability Science, models method and indicators, in Revista Critica de Ciências Sociais, 2011, Coimbra, Portugal

DA CUNHA, Gilberto Rocca et al. El Nino/ La Nina- South Oscillation and its Impacts on Brazilian Agriculture: Facts, speculation and application in Revista Plantio directo, V.123, Aldeia Norte Editora, Brazil, 2011, 7p

FROM PEACE, Adriano Rolim. Applied Hydrology, Brazil, 2004, 138p

DE ALCANTÂRA, V S, Vulnerability Socio-environmental in the Macroregion of Costa Verde, IBGE, Rio de Janeiro, 2012, 127p.

DE Almeida, A. Betâmio. Risk and Uncertainty Management: Concept and underlying philosophy, University of Lisbon, Lisbon, Portugal, 2005, 29p

DE ALMEIDA, Lutiane Queiroz. Socio Environmental Vulnerability of Urban Rivers: Maranguapinho River Basin, metropolitan region, Unesp, São Paulo, Brazil, 2010, 247p

DE ANDRADE, Marina Marconi & LAKATOS, Eva Maria Metodologias de Investigação Cientifica, 6th Edition, Atlas, São Paulo, 2007, 301p

DE CARVALHO, Daniel Fonseca & DA SILVA, Leonardo Duarte Batista. Hidrologia, Brazil, 2006, 115p

DE CASTRO, Augusto et al. Environmental Risk Management, Brazil, 2003

DE LIMA JUNIOR, Joaquim Alves et al. Estudo do Processo de Salinização Para Indicar Medidas de Prevenção de Solos Salinos, V.6 UNESP. Brazil, 2010

DEVIS, Claire. Climate Risk and Vulnerability: A Handbook for Southern Africa, CSIR Publisher, Pretoria, South Africa, 2011, 92p

National Directorate of Water (DNA), Southern Africa: River Basins, 1999

National Water Board. Hydrogeological map of Mozambique, scale 1:8000000, The Hague, Netherlands, 1992

DO AMARAL, Erico Hoff et al. Metodologia Para Calculacao do Risco Por Composição de Método: Programa de Pós-graduação em Informática (PPGI) Universidade Federal de Santa Maria (UFSM), Brazil, 2010

THE MUCHANGES, Aniceto. Mozambique, Landscapes and Natural Regions, Globo Typography, Lda, Mozambique, 1999, 162p

DOS SANTOS, António Marcos et al. Water Resources and Climate Change: Speeches, impacts and conflicts, Volume 51, Universidad de los Andes, Merida, Venezuela, 2010

DOS SANTOS, Ivair Augusto Alves. The Hard Drought in Mozambique, Maputo, 2016

DUTRA, Rita de Cássio et al. Global Vulnerability Indicators: Methodological proposal for studies and risk mapping in slope areas, ufsc editora, Brazil, 2014, 406p

Enginyeria Sense Fronteres et al, Funhalouro District Resource and Needs Studies: Funhalouro District Rural Water Pilot Project 2010-2011, Inhambane, Mozambique, 2010,106p

ESTARQUES, Maria. Natural Disasters in Africa, Lisbon, 2003

YOU WERE, Claudio Jesus de Oliveira. Socio-Environmental Risk and Vulnerability: Conceptual Aspects, V 1, IPARDES, Curitiba, Brazil, 2011, 79 p

FAO, The State of The World's Land and Water Resource for Food and Agriculture, Earthscan, New York, EUA, 2011

FEITELSON, Eran & CHENOWETH, Jonathan. Water Poverty: Toward a meaning indicator, Elsevier Editora, Jerusalem, Israel, 2002, 280p

FERREIRA, JOSE Manuel Vicente. The Management of Organizations in Times of Organizational Change: Portugal in the face of the III Industrial Revolution, Higher Institute of Social and Political Science, Lisbon, 1986

FRANKLIN, Jane. The Politic of Risk Society in: Review for the journal of forensic psychiatry, V. 10, Ucl, London, U.K, 1999.

RANGE, Rogério Guitierrez. Water Uses, Water Resource Management and Historical Complexity in Brazil: Study of the Paraíba do Sul River Basin, Brazilian Institute of Geography and Statistics, RJ, Brazil, 2009

GIL, António Carlos, Métodos e Técnicas de Pesquisa Social, 6th Edition, Atlas SA Editora 2006

GEICK, Peter H. Basic Water Requirements for Human Activities: Meeting basic needs, V.21, IWRA, Oakland, USA, 1996

Government of the District of Funhalouro, Plano Estrategico de Desenvolvimento Distrital- PEDD (2011-2015), Funhalouro, Mozambique, 2011, 90p

GONCALVES, Maria da Conceição et al. Características de Retenção de Agua no Solo para utilização na Rega das Culturas, INIAV Editora, Lisbon, Portugal, 2016, 69p

GONCALVES, Pedro. The Environment: A reflection on environmental problems, São Paulo, 2010

GPNCALVES, Maria da Conceição et al, Salinização do Solo: Causas e Consequências, Estacão Agrónoma Nacional de Portugal, Portugal

GUIGUER, Nilson & KOHNKE, Michael Wolfgang. Methods for Determination of Aquifer Vulnerability, Waterloo Hydrogeologic, Canada, 2016, 13p

HENSON, John et al. Environment science, population, Action Internation, Virginia 1998

HERCULANO, Lário Moisés Luís. Implementation of Alternative Sanitation Technologies as a Way to Guarantee Quality Water Quantity and Hygiene in the Mozambican Semiarid: Case of the district of Funhalouro, UFRGS, Porto Alegre, Brazil, 2012, 95p

HIPOLITO, João Reis & VAZ, Álvaro Carmo. Hydrology and Water Resources, Ist Editora, Lisbon, Portugal, 2011, 793p

HOEKSTRA, Arjen Y et al. Water Footprint Assessment Manual: Setting Global Standards, Earthscan Editora, Sao Paulo, Brazil, 2011, 183p

INAM, Manual for the Interpretation of Seasonal Forecasts, Proder Editora, Beira, Mozambique, 2002, 43p

INE, III General Census of Population and Housing 2007: Funhalouro District tabulation plan Mozambique, 2007

INIA. Mozambique Land Charter: Scale 1:1000000, 1994

ISDR, Terminology on Disaster Reduction, Switzerland, 2009

IWAMA, Allan Yu et al. Risk, Vulnerability and Adaptation to Climate Change: An interdisciplinary approach, Brazil, 2006

JACOBI, Pedro Roberto & GRANDISOL, Edson. Water and Sustainability challenges, perspectives and solutions, IEEUSP, São Paulo, 2017

KOBIYAMA, Masato et al. Recursos Hídricos e Saneamento, Organic Trading Editora, Brazil, 2008

KOBIYANA, Mosato et al. Natural Desert Prevention: Basic Concepts, Organic Trading, Curtiba, Brazil, 2006

LEITAO, Sanderson Alberto Medeiro. Water Scarcity in the City: Risk and Vulnerability in the context of Curitiba/Pr, UFPR, Curitiba, Brazil, 2009, 232p

LIMA, Angelo. Climate Change and Water Safety, WWF, Brazil, 2012

LIMON, Ramon Ruiz, Historia y Evolucion del Pensamento Cientifico, 2007

MAE. Profile of Funhalouro District Inhambane Province, 2005

MARSHALL, Samantha. The Water Crises in Kenya: Causes, Effects and solutions in: Global Majority E-Journal, V II, Bangladesh, 2011, 45p

MARTINS, Raul François RC. About the Strategy Concept, Instituto da defesa Nacional, Lisbon, 1983

MASSING, Atanasio Raimundo. Deforestation in Namaacha District and its Socio-Environmental Impacts - Case of Changalane Administrative Post, UP, Maputo, 2012, 60p

MENDES, José Manuel et al. Social vulnerability to Natural and Technological Hazards in Portugal, Revista critique de ciências sociais, Portugal, 2011

MICOA, Climate Change Vulnerability Assessment and Adaptation Strategies, Maputo, 2005

MICOA. Disaster Risk Management Capacity Assessment, Maputo, 2005

MICOA. National Climate Change Strategy 2013-2025, 2013.

MICOA. National Action Programme for Climate Change Adaptation (NAPA), 2007

MILLS, Landell, PORRAS, I.T. Silver bullet or fools gold? A global review of market for forest environment services and their impacts on the poor, IIED, London, 2002

Ministry of Agriculture, Sea, Environment and Land Management. Agriculture and Forestry Adaptation Strategies to Climate Change, Portugal, 2013

Ministry of Public Works. Water and Public Health, 2012

Ministry of Agriculture. Soil chart on the scale 1:1000000, Inia, 1994.

MORRISON, Jason et al. Water Scarcity & Climate Change: Growing risk for businesses & investors, Ceres, USA, 2009

MUVALE, Victor, Diário de Moçambique: Chigubo after three years of hunger and thirst, Maputo

NARVAES, William. Climate Change: The aprroach multidisciplinarity, 2nd Edition, UK, 2007, 367p

NAVARRO, António Fernando, Environmental Cost Analysis and Management, Brazil, 2004

NEPOMILUEVA, Daria. Water Scarcity Indexes: Water availability to satisfy human needs, Metropolia Editora, Helsinki, Filandia 33p

NÓBREGA, Jackson Silva et al. Study on the Viability of Cistern Use in Rural Settlement in Varzea-PB in Revista Verde de Agroecologia e Desenvolvimento V11, Pombal, Paraíba, Brazil, 2016, 5p

NUCCI, João Carlos. Environmental Quality and Urban Empowerment: A study and planning of the landscape applied to the district of santa Cecilia (MSP), 2 edition, Curitiba, São Paulo, Brazil, 150p

OECD. Disaster Risk Assessment and Risk Financing, Mexico, 2012

OLIVEIRA, Manuel Mendes. A Methodology for Calculating Surface Infiltration in a Sequential Water Balance Model Diário de Solos, LNEC Editora, Lisboa, Portugal, 2006

ORR, Stuart et al. Understanding Water Risk: A premier on the consequences of water scarcity for government and business, wwf, UK, 2009, 40p

OSORIO, Quelen da Silva. Natural Vulnerability of Aquifers and Groundwater Pollution Potential, UFSM, Rio Grande do Sul, Brazil, 2004, 136p

PAIVA, J. B. C & PAIVA, E. M. D. C. Hidrologia Aplicada a Gestão de Pequenas Bacias Hidrográfica, ABRH, Porto Alegre, Brazil, 2001

PEDROTTI, Aliceu et al. Causes and Consequences of Soil Salinization Process, UFSM, Brazil, 2015, 1324p

PEREIRA, Luís S. et al. Internation Hydrological Programme: Coping with water scarcity, UNESCO, Paris, 2002

PERREIRA, Cristina Lucas, Water Scarcity Assessment and its Use for Modelling the Water Resources Rate, Lisbon, 2017

Strategic Plan for District Development-PEDD (2011-2015). Funhalouro in the Leadership of Economic and Sustainable Development of the Community, 2011

UNDP. Sustainable Development Goals, USA, 2015

PRONASAR. Baseline Study on the State of Water Supply and Rural Sanitation: Report on the institutional capacity of Funhalouro district Inhambane province, Maputo, 2011.

RASKIN et al, Comprehensive Assessment of the Freshwater Resources of the World: Report of the Secretary General United Nations Economic and Social Council, Commission on Sustainable Development, Stockholm Environment Institute, Suecia, 1997

REICHERT, Jose Miguel et al. Soil Hydrology Water Availability and Agroecological Zoning, Brazil, 2011

HUMAN DEVELOPMENT REPORT: Water-Risk Scarcity and Associated Vulnerabilities, 2006

REPUBLIC OF MOCAMBIQUE, Profile of Funhalouro District, 5th Edition, Ministry of State Administration, Maputo, 2005

REPUBLIC OF MOCAMBIQUE, National Action Plan to Combat Drought and Desertification, Maputo, 2005

REUBOCA, André C et al. Águas Doces no Brasil, 2ª edition, instituto de estudos avançados da USP/ Academia Brasililiense de Ciências e Escritura Editora. São Paulo, 2002

REZENDE, Patrícia Soares. Methodology for Socio-Environmental Vulnerability Assessment: Study of the city of Paracatu (MG), UFU, Uberlândia, Brasil, 2016

RIJSBERMAN, Frank R. Water Scarcity: Facto r Fiction, IWMAI, Colombo, Sri Lanka, 2006, 14p

SAITO, Silva M. Natural Disasters: Basic Concept, Brazil, 2005

SALATI, Eneas et al. Relevant Environmental Issues, Brazil, 2006

SCOTT, Dianne et al. The Story of Water in Windhoek: A narrative approach to interpreting a transdiciplinary process in: Water, MDPI, Suíça, 2018, 1366p

SERRA, Carlos et al. 1 Monitoring Report on Good Governance and Environmental Resources in Mozambique, centre Terra Viva, Maputo, 2012, 327p

SIGENU, Kholisa. The Role of Rural Women in Mitigating Water Scarcity: A Dissertation Submitted in The Fulfiment of The requirement For the Master's Degree in Social Science, University Of Free State, Bloemfontein, South Africa, 2006, 114p

SOUZA, Katia Regina Goes. The Evolution of the concept of Natural and Social Sciences Risk, UERJ, Rio de Janeiro, Brazil, 2015, 43p

TROLEIS, Adriano Lima & DOS SANTOS, Ana Cláudia Ventura, Zonas semiáridas, 2nd Edition, UFRN Editora, Rio de Janeiro, 2011,

TUNDISI, José Galizia. Water Resources in the Future: Problems and Solutions, São Carlos, São Paulo, 2008, 16p

TUNDISI, José Galizia. Water Resources, International Institute of Ecology, São Paulo, Brazil, 2003, 13p

UN- Water. Nature-based solution for water management: Human Development Report, Italy, 2018

UN-ISDR, Living with Risk: A global review of disaster reduction initiative, Geneva, 2004

UP, Comissão de Revisão Curricular Central: Norma para publicação de trabalhos científicos na universidade pedagógica, Maputo,2004

GREEN, João Carlos. Forest Fire Hazard Assessment, University of Lisbon, Portugal, 2008

VERIATO, Mara Karinne Lopes et. Agua: Scarcity, crisis and perespectiva for 2050, GVAA, Brazil. 2015, 17p

VEYRET, Yvette et al. The Man as Aggressor and Victim of the Environment, São Paulo, 2007

VIANNA, Daniel de Berrito. Environmental Assessment and Risks in Contaminated Areas: A methodological proposal, Brazil, 2010

VIVAS, Eduardo & MAIA, Rodrigo. Scarcity and Drought Management Framing Climate Change, Portugal, 2010

VÖRÖSMARTY, C. J., et al. Global threats to human water security and river biodiversity. 2010.

WENG, C.C. Deser et al. El Nino and South Oscillation (ENSO): A Review, Springer Science Editora, USA, 2016, 106p

WHATELY, Marrusia & HERCOWITZ, Marcelo. Environmental Services: Know, value and care. Subsidy for the protection of São Paulo watersheds, social and environmental institute, São Paulo, Brazil, 2008

www.kunene.riverawarenesskit.com- consulted on 03 August 2018

XU, Hui & WU, May. Water Availability Indices- A literature review, Argonne Editora, Chicago, US, 2017, 31p

Appendix 1

AQUIFER VULNERABILITY IN THE FUNHALUR DISTRICT

As far as aquifer vulnerability is concerned, the God method (figure below) was used. According to Osório (2004:73), Foster evaluates the natural vulnerability of the aquifer and identifies the pollutant load, the interaction of which results in the preliminary characterization of risk areas.

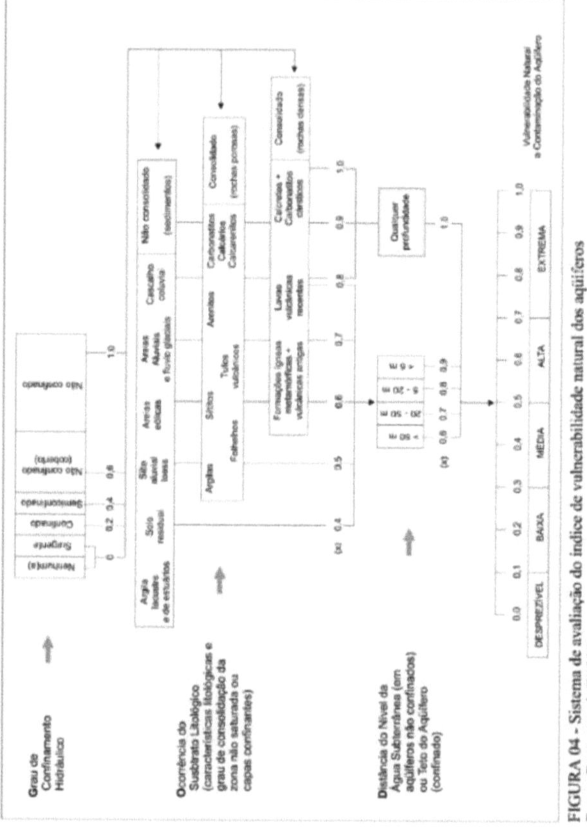

FIGURA 04 - Sistema de avaliação do índice de vulnerabilidade natural dos aqüíferos
Fonte: Foster et al. (2003)

Source: Osório (2004: 73)

The Gold method consists of:

1)   Identify the degree of hydraulic containment of the aquifer and mark it with a value between the scale of 0 and 1

2)   Specify the characteristics of the substrate above the saturated zone of the aquifer in terms of (a) degree of consolidation taking into account the likely presence or absence of crack permeability, (b) types of lithology considering indirectly the effective porosity, rock matrix permeability and moisture content in the unsaturated or retained zone and give it a value between a scale of 0,4 and 1,0

3)   Estimate the distance or depth of the water level in non-confined aquifers or depth from the ceiling of the first confined aquifer by assigning a value between the scale 0.6 to 1.0.
The final vulnerability index is the product among the results obtained for each of these parameters (1° x 2° x3°)

In the application of the GOD method, the analyzed parameters were given the following values:

Degree of hydraulic containment	G	Lithological substrate	O	Distance from water level	D	Vulnerability Index	Vulnerability class
Non-confined open	0,6	Carbonatids, limestones	0,9	60 meters to 70 meters	1,0	0,5	Top
Non-confined open	0,6	Areas, alluvial and glacial fluvial	0,7	60 meters to 70 meters	1,0	0,4	Media

The vulnerability of aquifer contamination to a polluting determinant in the district of Funhalouro varies from medium to high. In this context, one can give the example of a pollutant such as uncontrolled burning, deforestation and chemical fertilisers, which results in the loss of soil organic matter and later soil wear (erosion), which may give rise to sediment pollution, which according to Amorim (2011:17) results from the accumulation of soil particles and insoluble organic and inorganic chemicals in suspension and which, through the action of rain, is transported to the aquifer contaminating it or even interfering with the photosynthesis of plants.

FACULDADE DE CIÊNCIAS DA TERRA E AMBIENTE

CREDENCIAL/MESTRADO

Serve a presente para credenciar o/a Sr. _Atanásio Raimundo Maimuane_ ,Mestrundo do ___ ano, curso de _Gestão Ambiental_ ___________________________, desta Faculdade, para efeito de recolha de dados no (a) _Distrito de Funhalouro, província de Inhambane_

Subordinado ao tema _Escassez hídrica no distrito de Funhalouro._

No âmbito do trabalho de _Dissertação_

A Direcção da Faculdade agradece antecipadamente todo o apoio a prestar.

Maputo, _22_ de _Março_ de 2017

Appendix 3

INTERVIEW ITINERARY TO FUNHALOURO DISTRICT AUTHORITY

> This interview script was prepared as part of the Master's dissertation entitled Water Scarcity Risk in the district of Funhalouro with the aim of acquiring the degree of Master in Environmental Management by the Pedagogical University of Maputo

Interviewee Identification Data

Name

Date of Birth

Professional Training

One of the most worrying issues for the world in general and Mozambique in particular is the quantity and quality of water available for both human life and the economy and the district of Funhalouro characterized by the Semi-arid climate presents as one of the great challenges the supply of water both in qualitative and quantitative terms. This interview seeks the strategies adopted in the district of Funhalouro for the mitigation of social and environmental impacts of water scarcity.

INTERVIEW

1. What are the causes of water shortage in Funhalouro district?
2. What are the areas with the greatest and least water scarcity in the district?
3. What water supply infrastructure does the district have and what are the procedures for accessing water?
4. How much water is needed per day for the population?
5. What constraints does the district face in providing water to the public?
6. What is the current stage of environmental conservation in the district of Funhalouro

Order	Question	Answer	Person interviewed
01	What are the causes of water shortage in Funhalouro district?	Well, according to scientific investigations carried out by ESF in 2012 it is concluded that there are escapes of salts left in the rocks in the period of marine regression which contributes to the salty water, but on the other hand we find that the soil consisting of sandstone and conglomerates makes it difficult to open wells and water holes, needing for such equipment suitable for drilling, but the district government does not have the financial capacity to cover the costs of drilling operations at district level. In summary, there are two reasons for the water shortage in this district, one of an environmental nature and the other of an economic nature.	E.F.1
		It is said that one of the reasons for water scarcity in this district is due to the salinity of the water but it is not so because the drilling of water holes must go beyond the limestone layer and the drilling depth must vary from 60 meters to 100 meters, at these depths one finds water with less salt content than does not even appear to be Funhalouro water". It is necessary to drill and in the salt water towel should be done the insulation of pipes and put separators of the salt water with the fresh water as you continue with the drilling, but this costs a lot of money	E.F.2

02	What are the areas with the greatest and least water scarcity in the district?	The northern areas and one percent of the district are more vulnerable to water scarcity than the southern area due to the relief configuration and depth of the aquifers. However, in the southern zone what aggravates the lack of water most is the scarcity of rainfall because if rainfall occurred regularly the lakes and ponds would have water to supply the population	E.F. 1
		The southern zone of the district is less vulnerable to water scarcity when there is rain but the occurrence of rain in the district brings several health problems because in the rainy season there are more cases of acute diarrhoea, viruses, scabies and there is a record of deaths although in smaller numbers due to the consumption of water from lakes and ponds without proper treatment for human consumption.	E.F.3
03	What water supply infrastructure does the district have and what are the procedures for accessing water?	The District has manual water suction pumps although in small numbers in relation to the needs of the inhabitants of this district who are growing more and more, there are also wells and the PSAA (small water supply system) dominated by private operators, as well as recently built dams. Access to water in hand pumps and wells is free, nothing is paid for, but access to water from dams at the moment is still prohibited due to the lack of treatment of that water. As far as PSAA is concerned, the procedures for access to water depend on each operator.	E.F 1
		To have access to water in your residence from private operators it is necessary to pay the water contract worth 4000,00 Meticais and still pay the monthly consumption that is charged 50,00 Meticais for each m³, or buy 20 liters of water that cost 2,00 Meticais this in the administrative post of Funhalouro headquarters, Cupo and Mavume but for Mucuine, Manhiça and Tsenami the same amount of water costs 20,00 Meticais.	E.F.2
04	How much water is needed per day for the population?	The population needs at least 80 litres per day for domestic use in a household of 7 people.	E.F.2

05	What constraints does the district face in providing water to the public?	One of the major constraints in Funhalouro's water supply has to do with the constant breakdowns of some hand pumps for water collection	E.F.4
		The shortage and constant breakdown of existing hand pumps is one of the major constraints on water supply in this district and this is due to the ways in which they are managed, because once the hand pumps have been built they are handed over to community management for maintenance and repair in the event of breakdown, but the community does not have the financial capacity to maintain and repair the hand pumps for water collection, this constraint causes the population to move large distances in search of operational pumps for water. On the other hand, we do not have the financial capacity for the systematic maintenance of water pumps and there is a lack of technicians to carry out the necessary repairs on the pumps that break down.	E.F. 1
06	What is the current stage of environmental conservation in the district of Funhalouro?	As you may have seen along the road during your arrival in the village headquarters of the district, there is everywhere in the district the practice of burning and exploitation of forest resources for the manufacture of charcoal and use as construction materials, so the destruction of the environment is worrying, but we have a plan for reforestation although not using plant species native to the district, so it intends to plant approximately 1000 cashew seedlings, as well as make the population awareness to let minimize the use of burning in the opening of agricultural fields and hunting activity.	E.F. 1

Appendix 4: QUESTIONNAIRE

This survey was prepared within the framework of the Master's thesis entitled Risk of
Water scarcity in the district of Funhalouro for the purpose of acquiring master's degree
in Environmental Management by the Pedagogical University of
Maputo

Age ☐   Sex   M ☐   F ☐

Naturality ☐

Place of Residence ☐

Degree of Instruction   Professional Activity

Degree of Instruction		Professional Activity	
Can't read or write		Agriculture and Livestock	
Complete Basic Education		Industry	
Incomplete elementary school Secondary school		Trade	
High School		Student	
college education		Public Services	

What's your water source?	The place where you get the water is
Well ☐ Rio ☐ Lake ☐	Away ☐ Too far away ☐
Tap ☐ Fontenaria ☐	Near ☐ Very close ☐
Other source __________	At home ☐

Do you need to buy the water?	In case of YES what is the price of water?
Yes ☐ No ☐	__________ How many litres ☐

How many litres of water are spent per day

Less than 20L ☐   20L ☐   40L ☐   60L ☐   80L ☐

100 L ☐   150L ☐   300L ☐   Over 500 L ☐

The amount of water you buy in your Is source enough for your needs?	What is the quality of the water you get?

Yes [ ]   No [ ]

In case of No why?

Money [ ]   Shortage [ ]

Distance [ ]

Salobre [ ]   Very salty [ ]

Unsalted [ ][ ]   Potavel [ ]

Non-drinking [ ][ ]   Cloudy [ ]

What are your needs?
that consume the most water?

Domestic [ ]   Other purpose

Agriculture [ ]   ___________

Have you ever felt the lack of water?

Yes [ ]   No [ ]

What could be causing the lack of water?

Deforestation [ ]   Lack of government will [ ]   Wizards [ ]

Lack of Rain [ ]   I don't know. [ ]   Lack of money [ ]

How Water Management Works Over Time

Printed by Books on Demand GmbH, Norderstedt / Germany